Historical Studies
in the
Physical Sciences
5

Notice to Contributors

Historical Studies in the Physical Sciences, an annual publication issued by Princeton University Press, is devoted to articles on the history of the physical sciences from the eighteenth century to the present. The modern period has been selected since it holds especially challenging and timely problems, problems that so far have been little explored. An effort is made to bring together articles that expose new directions and methods of research in the history of the modern physical sciences. Consideration is given to the professional communities of physical scientists, to the internal developments and interrelationships of the physical sciences, to the relations of the physical to the biological and social sciences, and to the institutional settings and the cultural and social contexts of the physical sciences. Historiographic articles, essay reviews, and survey articles on the current state of scholarship are welcome in addition to the more customary types of articles.

All manuscripts should be accompanied by an additional carbon- or photocopy. Manuscripts should be typewritten and double-spaced on 8½″ X 11″ bond paper; wide margins should be allowed. No limit has been set on the length of manuscripts. Articles may include illustrations; these may be either glossy prints or directly reproducible line drawings. Articles may be submitted in foreign languages; if accepted, they will be published in English translation. Footnotes are to be double-spaced, numbered sequentially, and collected at the end of the manuscript. Contributors are referred to the *MLA Style Sheet* for detailed instructions on documentation and other stylistic matters. (*Historical Studies* departs from the MLA rules in setting book and journal volume numbers in italicized Arabic rather than Roman numerals.) All correspondence concerning editorial matters should be addressed to Russell McCormmach, Department of History of Science, Johns Hopkins University, Baltimore, Md. 21218.

Fifty free reprints accompany each article.

Historical Studies in the Physical Sciences incorporates *Chymia,* the history of chemistry annual.

The Physical Institute of the Zurich Polytechnic circa 1900. (*Festschrift zur Feier des fünfzigjährigen Bestehens des eidg. Polytechnikums* (Zurich, 1905), *2*, 336.)

Historical Studies in the Physical Sciences

RUSSELL McCORMMACH, *Editor*

Fifth Annual Volume

PRINCETON UNIVERSITY PRESS
PRINCETON, NEW JERSEY

Published by Princeton University Press, Princeton and London

ISBN: 0-691-08155-7
LCC: 77-75220

This book has been composed in IBM Selectric Aldine Roman

Printed in the United States of America
by Princeton University Press, Princeton, New Jersey

Editor's Foreword

A common feature of the articles in this fourth volume of *Historical Studies in the Physical Sciences* is their concern with modern physics in its relation to other scientific disciplines and to its philosophical and material context. The articles by Kuznetsov and Nye deal primarily with the interaction of physics and philosophy, Fox and Silliman with methodological and worldview considerations in physics, Culotta with the mutual relations of physics, philosophy, and biology, and Kohler with the hybrid discipline of physical chemistry. The first article in the volume is a multinational comparison of manpower, financing, and research productivity in physics around 1900. Forman, Heilbron, and Weart have developed categories for multinational comparison and with them have shown ways of picturing at a point in time the scale and rate of growth of the world's research activity in physics. Their geographical mapping of research capability in physics in 1900 will interest historians who are concerned with the multinationalism of the modern physics discipline; a first consideration in assessing the contributions of national physics groups is the material constraints on their research opportunities. In addition, their statistical approach will interest historians who wish to explore quantitative methods in their own work. Several articles in this and the previous volume of *Historical Studies* point to a larger range of possible contributions in the future. In keeping with this trend, in the second article I have chosen to discuss historical problems in physics around 1900.

Physics *circa* 1900

Personnel, Funding, and Productivity of the Academic Establishments

by Paul Forman, John L. Heilbron,

and Spencer Weart

Historical Studies in the Physical Sciences · 5

Contents

*The Bibliography contains full citations, which appear in the footnotes in abbreviated forms.

INTRODUCTION

We present here an inventory of the academic physics establishments maintained in the scientifically advanced countries at the turn of the century. Although most of the results are expressed quantitatively neither subject nor sources permit precision: the reader should constantly keep in mind the approximate character and devious construction of the figures in our tables. These figures, though approximate, can nonetheless be useful, not only to historians concerned with physics circa 1900, but to anyone seeking a benchmark against which the state and growth of physics in other eras can be measured. We do not here address the ultimate questions of the causes of growth (or retrenchment) of physics establishments, or the role of material factors in national scientific styles and achievements. But we are persuaded that historical generalizations and international comparisons brought forward in answer to these questions must prove compatible with—if they do not build upon—the quantitative data which we offer here, and which we hope others will improve and amplify.[1]

Our survey began as a preliminary to a study of the development of atomic physics from 1895 to 1930. That development is intriguing not only as intellectual history, but also because it took place de novo, recruiting its practitioners and acquiring its resources from the wider enterprise

[1]The principal precedent and parallel to our study is Karl Hufbauer's work on eighteenth-century German chemists, Hufbauer, *Hist. Stud. Phys. Sci., 3* (1972), 205–31 and *German Chem. Community (1700–1795)* (1970). Our results for a single discipline must ultimately be set in the context of the support of science and higher education as a whole. Essays toward the collection of such data for the period 1850 to 1914 have been made for Great Britain by Roy M. MacLeod and E.K. Andrews, "Selected Science Statistics" (1967), and for Germany by Frank Pfetsch, *Entwickl. der Wissenschaftspolitik* (1974).

of fin-de-siècle physics. We aimed to describe the growth of the field in terms of quantifiable sociological and economic parameters characterizing physics as a whole, as well as by reference to the appropriate theoretical and experimental advances. We found, however, that only one genus of the family of physicists, namely the academic, lent itself to detailed and uniform socioeconomic description: it was relatively homogeneous in composition, and homologous in structure from nation to nation, and its economy was well-documented in accessible published sources. Consequently we have specialized to academic physicists. The restriction has proved quite acceptable. The academic, by far the most numerous of the family of physicists in 1900, did most of the basic physical research of the time.

Although the data exist and are implicitly comparable, they must be coaxed from a wide range of sources—contemporary descriptions and surveys, publications of learned societies, reports of meetings, institutional and government documents, histories, biographies—which usually display the scant useful data they contain in categories different from ours and from one another's. Fortunately we have had the help of several able research assistants in winnowing this material.[2] With such aid, and by exercise of a little of the bibliographic resourcefulness proper to the historian, it is possible to generate for past times the sort of systematic, comprehensive, quantitative data which the social sciences require, but do not always provide, for their own use.

We lump our results in Tables I and II, which estimate, respectively, the annual world-wide investment in academic physics in 1900 and its rate of change in the following decade. For convenience we have used German marks which, in 1900, had roughly the same purchasing power as dollars do today.[3] Although we report statistics on all countries for which we

[2]Ted Bogacz, Michael Curtin, Judy Fox, John May, Michael Meo, Philip Pauly, Mary Pirruccello, Robert Seidel, and Neil Wasserman, supported through an NSF grant which we all gratefully acknowledge. In addition we are most obliged to Peter Heimann and Alan Chapman, and Gerald L'E. Turner for providing us with budgetary information about the Cambridge and Oxford laboratories, respectively; and to Joan Warnow and the American Institute of Physics (AIP) for making available to us their collection of unpublished histories of physics departments of American universities.

[3]Cf. the cost of housing, etc., in Krieg, *Bücher-Preise* (1953), p. 33. The equivalence of the 1900 mark and the 1973 dollar is very rough indeed: between 1900 and 1957 the purchasing power of the dollar fell by a factor of 3 or 4 (U.S., Bur. of Census, *Hist. Stat.* (1960), pp. 116–17, 125–26); in 1900 one got 4.2 marks to the dollar, whence a dollar in 1957 bought somewhat more than a mark in 1900. The equiva-

have sufficient data, we have directed our attention to the Big Four: France, Germany, the United Kingdom, and the United States. All historians would agree that the first three were qualitatively distinct from other physics-producing nations. Our tables show that the United States too belongs among the European powers with respect to numbers of practitioners and publications, and level of expenditures. Whether it also approached them on the intellectual plane cannot be decided by our numbers which, however, place this much controverted issue in a new light.[4]

The Tables' rubrics should be understood as follows.

A. *Number of Academic Physicists.* The first of our estimates is of the number of men occupying teaching positions explicitly devoted to physics (experimental, mathematical, and theoretical, but not technical, applied, industrial, or medical) in the higher schools (universities, technical institutes, normal schools) of the several nations. Needless to say the choice of higher schools involves some arbitrariness, especially in the smaller countries; we suspect, for example, that we have been too generous to Austria-Hungary, and have slighted Italy and the Netherlands. The United States presents special problems, but we are satisfied that we have dealt with them in a way which ensures comparability with the other leading countries.

The United States was first in total of academic physicists (Table I), followed by Germany, the United Kingdom, France, Austria-Hungary, Italy, Russia, Scandinavia, Switzerland, the Netherlands, Belgium, and Japan. The highest relative rate of increase of academic physicists (Table II) was sustained by Japan; the United States followed, with a relative rate half again as great as the fastest growing European establishments, the British

---

lents, in marks, of the chief currencies in 1900 were: Austrian krone, 0.84; Austrian florin, 1.68; Belgian, French, and Swiss franc, 0.82; Danish, Norwegian, and Swedish crown, 1.1; Dutch florin, 1.7; Finnish mark, 0.81; Italian lira, 1.3; Japanese yen, 2.1; British pound, 20.6; Russian ruble, 2.2; U.S. dollar, 4.2. France, Inst. nat'l. stat., *Annuaire stat., 58* (1951), 503*; *World Almanac and Encyclopedia* (1901), p. 175. For convenience we ignore fluctuations and use the exchange rates of 1900 throughout.

[4] For the issue see Reingold, "American Indifference" (1971), pp. 55–56; Slichter, cited by Dupree, *Sci. Fed. Gov.*, p. 300; and the interesting exchange between Carl Snyder and Simon Newcomb in *North Amer. Rev., 174* (1902), 59–72 and 145–58. Cf. Konen, *Reisebilder,* p. 56, a report of a trip to Pasadena in 1910: "One ought not to be deceived by well meaning articles in the newspapers which repeat the old view that Americans have money and institutes but no researchers or ideas. Perhaps this is still true in other fields, but in astrophysics and physics it has been out of date for a long time, and implies a fatal error. The Americans possess both, men and ideas, money and instruments; and they apply them with reckless energy." Konen, "Internat. Organization" (1911), p. 43, is also pertinent.

TABLE I.

Conspectus of Academic Physics circa 1900.

Faculty plus assistants	A. Number of academic physicists[a]		Expenditure (annual, 1000's of marks)						E. Productivity (physical papers)			
		per million of population[e]	B. Personal income[b]	C. Laboratory expenditures[c]	D. New plant[d]	Total: B + C + D	per physicist	per 10^8 marks national income[e]	Total (annual)	per physicist	per 10^4 marks expended	
Austria-Hungary	64	1.5	305	155	100	560	8.8	8.0				Austria-Hungary
Belgium	15	2.3	75			(150)[f]	10	3.8				Belgium
British Empire	131		950	315	770	2035	15.5		290	2.2	1.4	British Empire
U.K.	114	2.9	815	225	610	1650	14.5	4.6				U.K.
Other	17		35	90	160	385	22.5					Other
France	105	2.8	635	290	180	1105	10.5	4.3	260	2.5	2.4	France
Germany	145	2.9	765	385	340	1490	10.3	4.4	460	3.2	3.1	Germany
T.H.	36		150	85	100	335	9.3					T.H.
Univ.	109		615	300	240	1155	10.6					Univ.
Italy	63	1.8	260	180	80	520	8.3	3.4	90	1.4	1.7	Italy
Japan	8	0.2										Japan
Netherlands	21	4.1	95	90	20	205	9.8	7.1	(55)	2.6	2.7	Netherlands
Russia	35	0.3	160			(300)[f]	8.5	1.5				Russia
Scandinavia	29	2.3	110	90	45	245	8.5	4.1				Scandinavia
Switzerland	27	8.1	115	90	15	220	8.2	8.1				Switzerland
United States[g]	215	2.8	1430	660	900	2990	14.0	4.8	240	1.1	1.2[h]	United States[g]

[a]From Table A.1. "Faculty" signifies all ranks from professor to instructor and includes Privatdozent.

[b]Academic salaries and fees, calculated by multiplying the number of positions, Table A.4, by the average incomes of academic posts, Table B.1. We do not venture to estimate extramural professional and private income of the academic physicists. We have, however, added from Table C.4 income from prizes; only for France and Italy is this addition appreciable (16,000 and 9,300 marks, respectively). Salaries of nonacademic employees (janitors, mechanics, etc.) are included in "C", expenditure on operation of laboratory.

[c]From Table C.1.

[d]We have entered in Table I the annual average of the data in Table D.1 over the years 1895–1909 (i.e., 1902 ± 7, in order to allow for the delay between the funding and the completion of laboratories) augmented by the following sums (1000's of marks): Austria-Hungary, 25 (30 percent); United Kingdom, 100 (20 percent); other British Empire, 30 (20 percent); France, 40 (20 percent); German Univ., 20 (10 percent); German T.H., 10 (10 percent); Italy, 23 (40 percent); Netherlands, 10 (100 percent); Scandinavia, 10 (30 percent); Switzerland, 8 (100 percent); USA, 150 ($3/2 \cdot 20$ percent).

[e]Sources for income and population: Hoffmann and Müller, *Das deutsche Volkseinkommen* (1959), p. 14 (averaging over the figures c. 1900); U.S., Bur. Census, *Hist. Statistics* (1960), pp. 14, 139; Kindleberger, *Economic Growth* (1964), p. 337, whose figure for British national income is 8 percent larger than Deane and Cole, *British Economic Growth* (1967), pp. 166–67. We reduce Kindleberger's French national income, given in 1938 francs, to 1900 prices by dividing by 7.3, from France, Inst. Nat'l. Stat., *Annuaire statistique, 58* (1951), 515*. For populations in other countries we use *Cambridge Economic History, 6* (1966), 61, which gives figures for French, British and German populations 4 percent larger (Fr.), 7 percent smaller (U.K., Ger.) than the figures we use. Netherlands, Centraal bur. stat., *Stat. Yearbook* (1971), pp. 16, 248–52; *idem., Zestig jaren stat.* (1959), p. 102; Danmarks Statistik, *Stat. Årbog* (1972), p. 435; Norway, Stat. Sentralbyr., *Hist. Stat.* (1969), pp. 91, 528–29; Sweden, Stat. Centralbyr., *Hist. Stat.* (1960), p. 265; Italy, Istit. centr. stat., *Sommario di stat. storiche* (1968), p. 210. National incomes of Austria-Hungary, Belgium, and Switzerland are extrapolated from 1913 values given by Kuznets, *Encycl. Soc. Sci., 11* (1937), 206; that for Russia from the 1913 value given by Falkus, *Economica, 35* (1968), 52–73. U. Zwingli and E. Ducret, *Schw. Zeits. f. Volkswirt., 100* (1964), 367.

[f]For want of data regarding expenditures on laboratory operation and new plant we have estimated the total expenditure to be roughly twice the amount of salaries and fees; in the nine countries for which we have such data, salaries and fees lie in the range 45–55 percent of total expenditure. (Here and in the following tables parentheses signal our guesses.)

[g]Derived from our set of 21 leading schools using the multiplier 1.5, as explained in the text.

[h]Per 1.5×10^4 marks expended; see Table E.3.

TABLE II

Rate of Change of Table I, 1900–1910
(% of 1900 values per annum)

	A. Number of academic posts[a]		B. Personal income[b]	C. Laboratory expenditures[c]		D. New plant[d]	Total[e]
	Faculty	Assts.		Regular budget	Other		
Austria-Hungary	2.3	1.8	2.1	1.7		16	4.5
Belgium	1.1	1.6					
British Empire	2.3	9.0				8	(6)
U.K.	2.2	8.3	2.3	(6)	(20)	9	5.5
Other	3.3	14	(5)	(10)		5	(7)
France	0.7	0	1.8	(4)[f]	(10)	-3.2	1.8
Germany	3.4	3.8	4.8			2.5	4.9
T.H.	2.3	0	4.2			1	(3.7)
Univ.	3.8	5.1	5.0	6.2	(10)	3	5.2
Italy	1.9	1.1	2.1	3.2		-2	2.0
Japan	18						
Netherlands	3.0	2.5	(2.6)	(3)		0	(3)
Russia	-2.7						
Scandinavia	4.4	5.5	6.5	7.5		10	7.5
Switzerland	3.5	3.3	4.8	2		0	3.3
United States	7.1	6.5	9.5	15	(35)	6	10

[a]Taken directly from Table A.2. Note that the rate of change in column A refers to the number of *posts,* not *physicists,* so that the figures in this column do not exactly parallel those in column A of Table I.

[b]Compounded of the increase in posts (Table A.2) and the rise in salaries and fees of the various ranks in the several countries (section B.1.a).

[c]The increase in the regular budget is taken from Table C.2. As for "other" funds (see Table C.1), we assume that intramural grants and direct philanthropy rose at the same rate as the regular laboratory budget. To this we add the increase in extramural grants, Table C.4, plus the Carnegie, Commercy, and Solvay benefactions.

[d]We take from Table D.1 the difference between the totals for 1890–99 and 1905–14 as percentage of the total for 1895–1909.

[e]Pfetsch's data, *Entwickl.* (1973), yield a relative rate of increase of funding of science and higher education as a whole in Germany circa 1900 about 50 percent greater than our figure for physics. Likewise, M. Lenz, *Gesch. U. Berlin, 3* (1910), 525. MacLeod and Andrews' "Selected Science Statistics" (1967) suggest that in Britain too the growth of academic physics was relatively sluggish. Likewise MacLeod, *Hist. Journ., 14* (1971), 353–54.

[f]Estimated from the rate of change for all science, 4.9 percent (Table C.2), on the ground that chemistry and the biological sciences did somewhat better than physics.

and Scandinavian. By 1910 the British Empire had overtaken Germany, and the United Kingdom alone had probably done so by 1914. The number of academic physicists in other Western European countries grew more slowly, but substantially—except in France where the population of physicists, like that of the country as a whole, stagnated.

B. *Personal Income.* Salaries were annual amounts regularly given for instruction; in certain cases professors also received part of the fees charged students for instruction and examination. The income given is the sum of salaries, fees, and prizes given the men in column A.

C. *Laboratory Expenditures.* These funds include the regular laboratory budget (covering maintenance, wages of mechanics and custodians, recurrent expenditures for equipment and supplies), and special appropriations, extramural grants, and private philanthropy, expressed as annual averages over the years 1896 to 1905. All figures relate to the same set of higher schools underlying the figures of column A.

D. *New Plant.* The figures are averages of the amounts spent on building and equipping new laboratories and enlarging existing laboratories at the higher schools in the several nations.

The sums of the entries in the three funding categories (B,C,D) give the total direct annual investments in academic physics by the several nations. Even allowing for the substantial difference in price levels between Europe and America, the United States remains in front. Here, however, the United Kingdom slightly precedes Germany due to its high salaries and the very rapid growth of its redbrick institutions. France trails at some considerable distance, Italy and Austria-Hungary follow at about half the French expenditure, and Scandinavia, Switzerland, and the Netherlands fall together, down roughly by another factor of two.

E. *Productivity.* Our investment figures for academic physics call for comparable data on the output of physical research. We give therefore the number of research papers published in 1900 by academics in six leading nations, together with two measures of efficiency: rate of publication per man and per mark of total investment. In gross output Germany is far ahead. The British Empire is next, followed at roughly 10 percent intervals by France and the United States; Italy is far behind. By both productivity indices Germany retains a substantial lead, but the Netherlands and France now precede the British Empire. Italy retains her relatively low position, but in the United States research output per man or mark is lower still.

Some measure of the relative effort being made by the several nations is obtained by relating the personnel and funding of their academic physics establishments to their total populations and national incomes. The result is most interesting: within the precision of our data the number of academic physicists per million of population and the expenditure on academic physics per milliard of national income was the same in each of the Big Four in 1900. The differences between Germany, France, the United Kingdom, and the United States in their expenditure per physicist were, thus, essentially the differences between their per capita national incomes. This was, however, a fleeting conjunction; as Table II shows, the French were unwilling to try any harder while the British and Americans were increasing their commitments to physics far faster than their national populations and incomes. The Swiss and the Dutch were making extraordinary efforts in 1900, the Italians rather effete (the Austro-Hungarian figures are suspect); in the following decade, all lost ground relative to Germany, Britain, and the United States.

In the five following sections, which correspond to the five main rubrics of Table I, we show how Tables I and II have been constructed and provide extensive supplementary data. We believe that the national totals are accurate to better than 20 percent. The entries for individual schools are often only informed estimates in which, however, we have some confidence, since they have proved accurate in almost every case in which we subsequently found the desired information. The figures enclosed in parentheses, and all those for unbudgeted items, are but guesses.

A. NUMBER OF ACADEMIC PHYSICISTS

1. Number of Academic Posts

Table A.1 gives the number of faculty and assistants in physics laboring in the higher schools of the several nations in 1900; Table A.2 gives the growth in the number of such positions in the decade 1900–1910. In the cases of Austria and Germany we have counted positions in the Technische Hochschulen separately, for these institutions did not have the right to grant research degrees in physics (Ph.D.'s), and in some cases they had no research facilities.[5] Outside of Germany and Austria we have admitted

[5] The absence of research facilities in physics (in contrast to chemistry) at the T.H. Berlin was remarked upon by Robins, *Techn. School Building* (1887), p. 59, and is clear from the plans and plates in Berlin, T.H., *T.H. zu Berlin* (1903). The T.H. Breslau, opened in 1910 with very large physical-chemical and electrochemical institutes, had no physical institute at all, but relied upon the University's. *T.H. in Breslau* (1910), p. 5. The situation was similar at the T.H. Vienna: Jaeger (1915), p. 389. In Germany prior to the Weimar period students could "specialize" in physics, i.e., study it beyond their first two years, only at T.H. Munich and T.H. Dresden. Krüger,

only a handful of technical institutes, which we have treated as universities in our counts.[6]

In all European countries we have included every university explicitly so-called. If we followed the same procedure in the United States we would confront several hundred institutions, most of which would have but one or two posts in physics and no research program or facilities. Instead we admit the twenty highest in total annual expenditures for faculty salaries.[7] To these we have added Clark University, which, despite its small size, had significant standing in physics.[8] In drawing the line at a number close to that of the universities and colleges of the United Kingdom (30), France (21), and Germany (21), we have included all schools with significant graduate programs in physics.[9] The number of academic posts in the set of

---

Zs. für techn. Phys., 2 (1921), 113–21. Even the Zurich Polytechnicum (the ETH), the technical institute with the highest scientific reputation, had remarkably few advanced students in physics. Crew, "Diary, 1895." Of a sample of 93 German academic physicists active in 1900 (infra, Table B.2) only 8 percent attended a technical institute at any time, and about half of these graduated from a university.

[6]Our set of institutions is shown in Table A.4 except for Russia, where we take the Univs. of Charkov, Dorpat, Kazan, Kiev, Moscow, Odessa, St. Petersburg, and Warsaw, and the Polytechnical Schools of Riga, Moscow, and St. Petersburg. We have counted the free (Catholic) universities only in Belgium. Including them would raise our numbers of full professors in France by about 7, and of junior faculty in Italy by about 4; the increase of all other numbers would be only a few percent. Cf. Paul, *Societas, 1* (1971), 271–85, and Pelletier, *Branly* (1962). Note that we have treated all Italian royal universities as equals, although they were explicitly divided into two classes (initially the Univs. of Bologna, Naples, Padua, Palermo, Pavia, Pisa, Rome, Turin in class I, those of Cagliari, Catania, Genoa, Messina, Modena, Sassari, Siena in class II). The division tended to depress and impoverish the second class schools. Catania, U., *Storia* (1934), pp. 357–59; Genoa, U., *Università di Genova* (1923), pp. 83–96; Mor, *Storia* (1952), pp. 139–41, 182–83; Palermo, U., *Al ministero* (1899), pp. 11–12, 57–59.

[7]These are, in order of decreasing expenditures: Columbia, Harvard, Univs. of Chicago, Michigan, Yale, Cornell, Univs. of Illinois, Wisconsin, Pennsylvania, and California, Stanford, Princeton, MIT, U. Minnesota, Ohio State U., Univs. of Nebraska and Missouri, New York U., Northwestern, Johns Hopkins, and Clark. Carnegie Foundation, *Bull., 2* (1907), 5.

[8]Cf. A.G. Webster in *Clark U., 1889–1899,* pp. 90–98.

[9]That the number of active centers of physical research was about the same in the United States, Germany, and Britain also appears from subscriptions to the Royal Society's *International Catalogue of Scientific Literature:* in 1906 the three nations received, respectively, 17, 14, and 18 subscriptions to the volumes for physics, as against 62, 44, and 29 subscriptions to the *Catalogue* as a whole. Schuster, *Nature, 74* (1906), 234. There were however a number of other institutions which had granted at least one Ph.D. in physics prior to 1917: Bryn Mawr College, U. of Cincinnati, George Washington U., U. Iowa, Stevens Inst. of Tech., Vanderbilt U. Cf. Kevles, *Phys. in America* (1964), pp. 309–13.

TABLE A.1

Number of Academic Physicists in 1900[a]

	Faculty						
	Senior	Junior	Privat-dozenten	Total	Assistants[b]	Research affiliates[c]	Total[c]
Austria-Hungary	26	9	13	48	(16)	(15)	79
Belgium	7	2	0	9	6	(2)	17
British Empire	40	34	12[d]	86	45[e]	(40)	171
U.K.	32	32	12	76	38	(30)	144
Other	8	2	0	10	7	(10)	27
France	27	26	0	53	52	(40)	145
Germany	38	31	34	103	42	(90)	235
T.H.	11	6	9	26	10	(10)	46
Univ.	27	25	25	77	32	(80)	189
Italy	18	7	18	43	20	(10)	73
Japan	4	2	0	6	(2)	(3)	11
Netherlands	8	1	1	10	11	(10)	31
Russia	10	8	12	30	5	(5)	40
Scandinavia	11	7	0	18	11	(5)	34
Switzerland	11	2	4	17	10	(20)	47
United States	32	31	36[f]	99	46	(50)	195
Total	232	160	130	522	266	290	1078

[a]Posts from Table A.4, corrected for *cumul,* except that Russian numbers are from *Minerva: Jahrb.* Counts from *Minerva: Jahrb.* generally fall only slightly short of the fuller data of Table A.4, except for the U.S., where it considerably underestimates the lower ranks.

[b]Assistants: from Table A.4. We have reduced the numbers of assistantships by half the numbers of Privatdozenten, as described in the text.

[c]Students and other affiliates of academic institutions engaged in research. These figures, as the parentheses are meant to indicate, are our guesses, informed, however, by the data on "advanced students" in Table A.4 and on postgraduate degrees in Table A.3.

[d]Fellows (estimate). Apparent discrepancy with Table A.4 due to *cumul.*

[e]Assistant lecturers and demonstrators salaried at less than £ 125 per annum.

[f]Instructors.

TABLE A.2

Growth of Physics Posts circa 1900[a]

	1890–1900		1900–1910				
	Total faculty	Assts.	Senior faculty	Junior faculty	Privat-dozenten	Total faculty	Assts.
Austria-Hungary		(0)	5	-1	7	11	4
T.H.			1	1	2	4	2
Univ.			4	-2	5	7	2
Belgium		(0)	-2	3		1	1
British Empire			-2	17	7[e]	22	47
U.K.		(20)	-1	13	7[e]	19	37
Other			-1	4	0	3	10
France	12[b]	20[b]	5	-1		4	0
Germany		(20)	11	-1	25	35	23
T.H.			3	0	3	6	0
Univ.	-3[c]		8	-1	22	29	23
Italy		4[d]	4	-3	7	8	3
Japan			3	8		11	
Netherlands		(1)	1	2	0	3	3
Russia		(1)	-1	-4	-3	-8	
Scandinavia		(0)	4	4	0	8	6
Switzerland			1	2	3	6	4
United States		(11)	19	21	30[e]	70	30
TOTAL		(80)	48	47	76	171	120

[a]Sources and basis for 1900–1910 as in Table A.1. We have not included schools founded or upgraded in this period, notably: T.H. Danzig, T.H. Breslau, T.H. Delft, U. Reading, U. Southampton. The 1890–1900 increase in the number of assistants is estimated, except as noted below, from Weinberg, *Phys. dans 206 lab.* (1902), who gives data for c. a fifth of our institutions.

[b]The growth 1887–1900, using France, Min. instr. publ., *Statistique, 3* (1889), *4* (1900); for posts established after 1900 see *idem., Recueil des lois, 6* (1909), 571, 633, 928–29, 1026.

[c]For 1890–1900 we use Ferber, *Lehrkörper, 1865–1954* (1956), p. 197; his total figures for physicists are slightly less than ours.

[d]For 1890–1900 we use the growth for 1891–1900 from Italy, Min. pubbl. istr., *Annuario* (1892, 1901).

[e]Privatdozenten: Fellows in the British Empire, instructors in U.S.

twenty-one American universities is ninety-nine. We note that of the 104 American academic physicists identified as prominent by Cattell in 1902,[10] only sixty-seven belonged to schools in our set. If therefore we increase the number of physics posts in our set by 50 percent, we should have a good estimate of the total of productive academic physicists in the United States c. 1900. We assume that the same multiplier will transform the physics budgets of the set into a reasonable figure for all American expenditures for physics. In Table I we have accordingly entered 1.5 times the American entries in Tables A.1 and C.1. The territories and dominions of the British Empire present a similar problem; we have admitted seven institutions (three in Canada, four in Australia) as on a par with those in the United Kingdom.

We have divided the "faculty" of Table A.1 into three categories. The first, the full professor, the chairholder and director of the institute, needs no further identification. By "junior faculty" we mean a miscellany of tenured and nontenured academics: assistant professors, lecturers, readers, and senior demonstrators in England; professeurs adjoints and suppléants, maîtres de conférences, chargés de cours in France; assistant, adjunct, and associate professors in the United States; ausserordentliche Professoren in Germany.[11]

The third class of faculty, the Privatdozent, is only with some impropriety said to have a "post." Strictly speaking he received no salary and the numbers of Privatdozenten were not formally limited. In practice, however, he occasionally (and his Italian counterpart always) received a small stipend,[12] and he required the acquiesence of the Ordinarii to gain entry into university teaching (Habilitation); hence the possibility of Privatdozent "posts." (For convenience we give the number of instructors in the United States and Fellows in the United Kingdom in the column for Privatdozenten, although instructors and Fellows were more nearly junior faculty.)

[10]Cattell, *Amer. Men of Sci.* (1906), passim; *ibid.* (1910), p. 583. Cattell's list includes many applied physicists but excludes junior faculty who had yet to make a reputation. Kevles, *Phys. in America* (1964), pp. 303–08, deems 92 American physicists of 1895 to have been distinguished, and so agrees well with Cattell.

[11]There is difficulty here, for some (the nichtetatsmässige) ausserordentliche Professoren held the title only, not the post, and consequently were not salaried. Note also (1) the Dozenten at the Technischen Hochschulen, in contrast to the Privatdozenten, were salaried (Damm, *T.H. Preussens* [1909], p. 264) and (2) the rank of extraordinary professor was abolished in Sweden in 1909, when all Extraordinariats became professorships (Weibull, *Historia, 4* [1968], 182–92).

[12]Infra, Table B.1.

By "assistant" we mean the lowest rank of academically trained laboratory or teaching help: the Assistenten, préparateurs, répétiteurs, chefs des travaux pratiques, demonstrators, and tutors and assistants in physics common in American universities. The figures for all countries but Russia in Tables A.1 and A.2 come from Table A.4, which displays our counts school by school; in the transition some posts may appear to have been lost, for we have corrected for double posts which customarily went together. In Germany, for example, an assistantship often—we have estimated 50 percent of the time—accompanied the position of Privatdozent.[13]

One will gather from Table A.1 the useful result that, on the average, there were one junior faculty or Privatdozent and somewhat more than one assistant for each professor of physics in 1900. The rule holds for all countries, despite differences in organization and structure of authority: discrepancies in staffing (as well as in level of funding) must be sought not between nations, but between large and small establishments. Outside the main centers—in the Münsters, Sheffields, Dijons, Sienas—the physics department might consist of two or three members. But in Cambridge, at the Cavendish laboratory, J. J. Thomson directed four lecturers, four demonstrators, three assistants, and twelve research students.[14] At Berlin Emil Warburg presided over three chairless faculty, eight Privatdozenten, eight assistants, and a score of research students.[15] And in Paris a like complement of assistants and researchers worked in the Sorbonne laboratories directed by Gabriel Lippmann and Edmond Bouty.[16]

Table A.3, which gives the numbers of students and their rates of change, taken together with Table A.2, shows that the ratios of the number of advanced students,[17] assistants (both academic and technical), and junior faculty to the number of chairs were increasing substantially in 1900.

The instructional staff rose largely because university physics had be-

[13]Infra, section B, note 27. The doubling of posts also occurred in the upper ranks, especially in France and Italy.

[14]Cambridge, U., *Hist. Register* (1917); Cavendish Lab., *History* (1910), pp. 82, 89, 221, 229–80.

[15]Lenz, *Gesch. U. Berlin, 3* (1910), 278 ff., 446; Küchler, "Phys. Labs. in Germany" (1906), p. 192; Berlin, U., *Chronik.*

[16]Weinberg, *Phys. dans 206 lab.* (1902).

[17]The growth of the research students at some major centers may be noticed: in 1900, there were a dozen each at the Cavendish, the Sorbonne, and the University of Vienna, 6 at Leipzig, 30 at Berlin, 10 at Harvard; the Cavendish had 30 in 1910, Leipzig 18 in 1908, Harvard 14 in 1905. Cavendish Lab., *History* (1910), pp. 9, 99, 101, 221; Küchler, "Phys. Labs. in Germany" (1906), p. 192; Leipzig, U., *Festschrift, 4*, Pt. 2 (1909), 57; Harvard U., *President's Report* (1899-1900), and (1904–05).

TABLE A.3

Students and Degrees in Physics
(with rate of change, Δ, number per annum)

	Postgraduate degrees[a]		"Graduate" students[b]		First degrees[c]		Undergraduate "majors"[d]		Students in laboratory courses[e]		Students in lecture courses[f]	
	1900	Δ	1900	Δ	1900	Δ	1900	Δ	1900	Δ	1900	Δ
Austria-Hungary: Univ.	1	+							400	35		
British Empire: U.K.			30		100	0	200	0	2500	50	3000	
France	6	0.1			60	5			1600	100	4000	
Germany	5	+	18		30	3	130	20	2500	190	5000	150
T.H.			3				10		1000	90	2000	0
Univ.			15		30	3	120		1500	100	3000	150
Italy					15	0	145	-5	600	13		
Japan	1	0.1			11	0.5	29	3				
Scandinavia	3	0.2							225	100	750	200
United States	15	1.0	200	-10	75	-5	150	-10	4000	220	8000	0

[a]Post-graduate degrees:

Austria: There were 3 Habilitationen in physics at U. Vienna in the 1870's, 5 in the 1880's, and 6 in the 1890's. Vienna, U., Akad. Senat, *Geschichte* (1898), pp. 291–92.

British Empire: Our guess of the number of candidates for honors degrees in physics in the United Kingdom.

Germany: Doktoranden and candidates for secondary school teaching certificate. In Prussia in 1902 there were 15 qualified

France: Dr.-ès-Sciences. Rate of change is for 1900–10; for 1890–1900 r. of c. is 0.2. France, Min. instr. publ., *Cat. des Thèses* (1890–1919).

Germany: Ferber, *Lehrkörper* (1956), gives 87 Habilitationen in physics in 1890–1909.

Japan: "Hakushi." See: Tokyo Imp. U., *Calendar* (1899–1900), pp. 27–30, 200–06, and note d.

Scandinavia: Doctorate. See Table A.4, note e.

U.S.: Doctorates, as compiled from *Science,* 1898–1917, by Kevles, *Phys. in America* (1964), p. 294.

[b]"Graduate" students:

British Empire: Our guess of the number of research students in physics in the United Kingdom.

Germany: Our guess of the number of students between Promotion and Habilitation who were not assistants.

U.S.: Graduate students from Table A.4 (under "Advanced students"), increased by 1/3 to allow for students at schools other than those in our set of 21.

[c]First degrees:

British Empire: Our guess of the number of honors B.A. and B. Sc. degrees in physics and natural philosophy in the United Kingdom.

France: Certificats de physique, from France, Min. ed. nat., personal communication (13 April 1973).

Germany: Our guess of the number of doctorates (Promotionen) in physics.

Italy: "Laureati" in physics. The annual average 1905–11 was 15 with no clear trend. Cf. Ferraris, *Annali di statistica, 6* (1913), x, 87; *Riforma sociale, 17* (1907), 738; R. istituto veneto di scienze, *Atti,* ser. 8, *13,* Pt. 2 (1910), 219.

Japan: "Graduates" from Tokyo (10) and Kyoto (1) Imperial Universities. See note d.

[d]Undergraduate "majors":

candidates with mathematics and physics as major subject; in 1904, 18; 1906, 26; 1908, 54; 1910, 111; 1912, 148. Prussia, Stat. Landesamt, *Stat. Jahrb., 11* (1914), 419.

Italy: "Inscritti per laurea in fisica," source as in note c.

Japan: "Undergraduates" at Tokyo (26) and Kyoto (3) Imperial Universities. Average age at "graduation" was 27 in 1911. Tokyo, Imp. U., *A Survey, 1913–14,* tables following p. 8 of supplement "Summary of School Expenses"; Japan, Imp. Min. Ed., *Annual Report, 28* (1900), 81, 91; *37* (1909–10), 197, 216; Kikuchi, *Japanese Education* (1909), p. 368; Tokyo Imp. U., *Calendar* (1909–10), appended tables.

U.S.: A guess for the 21 institutions in our set based upon data for Chicago (24), Cornell (10), MIT (10), suggesting that the number of physics majors was not much greater than the number of graduate students and was certainly not increasing before c. 1905. Cornell U., *President's Report* (1906–07), pp. cxii–iii; MIT, *Reports* (1900), p. 52.

[e]Students in laboratory courses:

Estimated on the basis of data from Weinberg, *Phys. dans 206 lab.* (1902), detailed in Table A.4. U.S. figures refer to the set of 21 institutions listed there. Weinberg's data imply: Belgium: 600, $\Delta = 30$; Netherlands: 400, $\Delta = 8$; Russia: 500, $\Delta = 15$; Switzerland: 550, $\Delta = 0$.

[f]Students in lecture courses:

Estimated on the basis of data in Table A.4. The figures for Germany are the number of students in the main introductory course in experimental physics; those for the U.S. are total enrollments in all undergraduate physics courses; those for Scandinavia are total enrollments for all physics courses in all schools except Stockholm Högskola. The large increases in Scandinavia are owing chiefly to the introduction of laboratory courses into Swedish secondary schools in 1905, and the consequent demand for trained teachers. Leide, *Fysiska institutionen* (1968), pp. 134–35.

TABLE A.4

Posts and Students in Physics Institutes

	Posts[a]										Students			
	1900					1910					Advanced, 1900[e]	In laboratory courses		In Lecture Courses, 1900
	Professor[b]	Jr. faculty[b]	Privatdoz.[b]	Assistant[c]	Employee[d]	Professor[b]	Jr. faculty[b]	Privatdoz.[b]	Assistant[c]	Employee[d]		1890	1900	
Austria-Hungary	26	9	13	(22)	(26)	31	8	(20)	(26)	(28)				
Tech. Hochsch.	7	2	3	(3)	(3)	8	3	5	(5)	(5)				
Brünn	1	1	0	1		1	0	2						
Budapest	1	0	0	(0)		1	0	1						
Graz	1	0	1	1		2	0	0						
Lemberg	1	0	0	(0)		1	1	0						
Prag Dtsch.	1	0	0	(0)		1	0	0						
Prag Böhm.	0	1	0	(0)		0	1	0						
Vienna	2	0	2	1		2	1	(2)						
Universities	19	7	10	(19)	(23)	23	5	15	(21)	(23)				
Agram	1	0	0			1	0	0						
Budapest	2	1	0			3	0	0						
Czernowitz	2	0	0	1	1 + 0	2	0	0	1					
Graz	2	1	1	1	2 + 2	2	0	1	1	0 + 3	(2)	5	24	
Innsbruck	2	0	3			2	1	2	2	1 + 1			17	
Klausenburg	2	0	0			2	0	1						
Krakau	1	1	1	2		2	1	0				16	51	
Lemberg	1	1	0			2	0	0						
Prag Dtsch.	2	1	1	2	1 + 2	2	0	0	3	1 + 2				
Prag Böhm.	1	1	0	2		2	1	2				14	71	
Vienna	3	1	4	5	2 + 3	3	2	9	5	2 + 3	(9)	12	36	

Belgium	7	2	0	6		5	5	0	(7)					
Brussels U. Libre	1	1	0	2		0	2	0					40	
Ghent U.	2	1	0	1		2	1	0				33	106	
Liège U.	2	0	0	3	0+	1	1	0	3	1+			300	
Louvain U. Cath.	2	0	0	0	1	2	1	0	0	1				
British Empire	40	38	(19)	52	(53)	38	55	(26)	(99)					
United Kingdom	32	36	(19)	45	(45)	31	49	(26)	(82)	(65)				
Aberdeen	1	1		1		1	2		2		0		17	106
Aberystwyth	1	1		1		1	1		(1)	(1)		16	61	
Bangor	1	0		2		1	1		1	(1)				46
Belfast Queen's	1	0		3		1	0						68	104
Birmingham Mason	1	1		1		1	2		3	(3)			76	
Bristol	1	1		1		1	1		2	(1)				263
Cambr. Cavendish	2	8	(11)	5	4	2	9	(14)		(6)	12		300	
Cardiff	2	0		2		2	0		(2)	(1)				
Cork	1	0		(0)		1	0							64
Dublin Trinity	2	0	(1)	3		2	1	(1)	4			38	39	100
Dundee	1	0		1		1	0		1	(2)				
Edinburgh	1	2		1	2+	1	2		4	2+	4		31	173
Galway Queen's	1	0		1		1	0							58
Glasgow	1	3		1		1	4		5	(3)				
Leeds Yorkshire	1	1		2	1+	1	(2)		3	2+		51	141	180
Liverpool	1	0	(1)	1	3½	1	2	(2)	4	(4)				
London U. Coll.	1	2		1		1	2	(1)	1	(3)			80	
Lond. King's Coll.	2	2		0		1	3		1	(2)			80	
Lond. R. Coll. Sci.	2	2		7		2	2	(1)	13		1		116	83
Manchester Owens	2	1	(1)	3		1	1	(2)	4	(3)	4		110	160
Newcastle Durham	1	2		(1)		1	2		3	(2)				
Nottingham	1	3		2	1	1	4		3	(1)		50	180	
Oxford	2	5	(5)	4	(3)	3	6	(6)	7	(5)			50	
St. Andrews	1	0		1		1	1		1	(2)				97
Sheffield	1	1		0		1	(1)		3	(2)				

(Footnotes and sources on pages 25–28.)

TABLE A.4 (*continued*)

	Posts[a]										Students			
	1900					1910					Advanced, 1900[e]	In laboratory courses		In Lecture Courses, 1900
	Professor[b]	Jr. faculty[b]	Privatdoz.[b]	Assistant[c]	Employee[d]	Professor[b]	Jr. faculty[b]	Privatdoz.[b]	Assistant[c]	Employee[d]		1890	1900	
Other	8	2	(0)	7	(8)	7	6	(0)	17					
Adelaide U.	1	1		0	1 +	1	(0)		1	1 +				
Dalhousie U.	1	0		0		1	(0)		1					
McGill U.	2	0		3	2 + 1	1	3		4				250	
Melbourne U.	1	0		1		1	1		2					
New Zealand U.	1	0		0		1	0		1					
Sydney U.	1	0		1		1	0		3					
Toronto U.	1	1		2	2 +	1	2		5		2		250	250
France	27	27		53		32	26		(53)		(40)			
Besançon	1	0		2		1	0					10	37	
Bordeaux	2	2		5		3	2							
Caen	0	2		1		1	1					12	35	
Clermont-Ferrand	1	1		2		1	1							
Dijon	1	1		1		2	0						(80)	
Grenoble	1	1		2		1	2						45	
Lille	1	2		3		2	3					24	86	200
Lyon	1	2		4		1	2		4					
Marseille	2	1		3	1/5 +	1	1		3				74	120
Montpellier	2	0		2	2 + 1	1	2		2					
Nancy	2	1		4		3	0							
Paris Univ.	4	6		12	6	5	7				(30)		260	
Paris Coll. France	2	1		1		2	0		1					
Paris Éc. Normale	0	3		3	2	0	3				(4)		45	

Paris Éc. Polyt.	2	1		6		2	1							700
Paris Éc. Phys. & Ch.	1	0		0	+ ½	1	0						45	
Poitiers	1	1		1		2	0							
Rennes	1	1		2		1	1						80	
Toulouse	2	1		2		2	1		1			19	127	
Germany	38	31	34	60	(58)	49	30	60	(83)	(78)				
Tech. Hochsch.	11	6	9	15	(18)	14	6	12	(15)	(23)	(10)			
Aachen	1	1	1	1		1	1	2						
Berlin	1	3	2	2		2	1	2				17		500
Braunschweig	1	0	1	1	0	2	0	2	1	0 + 1		20	30	
Darmstadt	1	1	1	3	2 + 1	1	1	1	2	2 + 2		29	311	200
Dresden	2	0	1	(2)		3	0	1						
Hannover	1	1	0	1		1	1	0	1					
Karlsruhe	2	0	1	2	2 + 2	2	0	1	2	2 + 2				500
Munich	1	0	2	2		1	1	2				45	290	
Stuttgart	1	0	0	1	1 +	1	1	1	1	1 +		6	31	250
Universities	27	25	25	45	(40)	35	24	47	68	(55)	(120)			
Berlin	3	3	8	5	1 + 4	4	2	12	8	2 + 4	(20)		70	400
Bonn	1	1	2	1		1	1	2	4		5		27	
Breslau	1	1	1	2	0 + 1	2	1	2	4	1 + 2	(2)	2	35	120
Erlangen	1	1	0	1	1 + 1	1	1	0	1	1 + 1			85	140
Freiburg i. Br.	1	1	1	2	1 + 1	1	2	1	2	1 + 3	(3)	29	73	180
Giessen	1	1	0	2	1 + 1	2	1	3	3	1 + 1	3			80
Göttingen	2	2	3	3	½ + 1	2	0	2	4	0 + 1			106	
Greifswald	1	2	1	2	0 + 1	1	2	2	2	0 + 1			57	120
Halle	1	1	0	1	1 + 1	1	1	1	3	1 + 1	6		40	
Heidelberg	1	2	2	4		1	2	3	(4)				100	
Jena	1	2	1	1	1 + 0	1	3	0	3	1 + 0			15	
Kiel	2	0	0	(2)		2	1	0	(2)		6		35	110
Königsberg	2	0	0	2	(3)	2	0	0	(2)				14	
Leipzig	2	2	2	3	1 + 1	3	1	5	8	2 + 3	6	46	116	340
Marburg	1	1	0	2	2	1	1	2	4	1 + 2	(3)		25	150
Munich	1	2	2	3	3 + 1	3	0	6	4	4 + (2)	(10)		60	
Münster	1	0	1	1		2	1	1	2				42	70
Rostock	1	1	1	2	0 + 1	1	1	1	1	1 + 1			10	

(Footnotes and sources on pages 25–28.)

TABLE A.4 (*continued*)

	Posts[a]										Students			
	1900					1910						In laboratory courses		
	Professor[b]	Jr. faculty[b]	Privatdoz.[b]	Assistant[c]	Employee[d]	Professor[b]	Jr. faculty[b]	Privatdoz.[b]	Assistant[c]	Employee[d]	Advanced, 1900[e]	1890	1900	In Lecture Courses, 1900
Strassburg	1	1	0	3	1 + 2	1	1	1	3	1 + 2				
Tübingen	1	1	0	1	1	2	1	2	2	1 + 2		11	30	75
Würzburg	1	0	0	2	0 + 1	1	1	2	2	1 + 1				
Italy	18	7	18	28	41	22	3½	25	31	50	144			
Bologna U.	1	1	0	4	2 + 1	1	1	0	2	4 + 2	36			
Cagliari U.	1	0	0	1	0 + 1	1	½	0	2	0 + 1	0			
Catania U.	1½	0	0	3	1 + 1	1½	0	1	2	1 + 1	5			
Genoa U.	1½	0	0	1	1 + 1	1½	0	0	2	1 + 1	7			
Messina U.	½	1	1	1	1 + 1	1	½	0	1	1 + 1	11			
Modena U.	1	0	0	1	1 + 1	1	0	2	1	1 + 1	2	3	11	
Naples U.	2	0	5	1	2 + 1	2	0	7	3	1 + 3	25			
Padua U.	1½	0	2	3	2 + 2	1½	0	2	3	2 + 2	6		35	
Palermo U.	1½	0	0	2	2 + 1	1	½	1	2	1 + 1	3			
Parma U.	1	0	0	1	2 + 1	1	0	0	1	2 + 1	0	6	10	
Pavia U.	1	1	0	2	1 + 1	2	0	0	2	1 + 1	7			
Pisa U.	1½	0	3	1	1 + 2	1½	0	3	2	3 + 3	13	8	32	
Rome U.	1	1½	4	4	2 + 3	3	0	6	4	4 + 4	14	84	84	
Sassari U.	0	1	0	1	0 + 1	0	1	0	1	0 + 1				
Siena U.	1	1	0	0	1 + 2	1	0	0	1	1 + 1				
Turin U.	1	½	3	2	1 + 1	2	0	3	2	1 + 2	15	28	45	
Japan	4	2		(2)		7	10							
Kyoto Imp. U.	2	1				3	4							
Tokyo Imp. U.	2	1				4	6							

Netherlands	8	1	1	12	(13)	9	3	1	15	(17)				
Amsterdam U.	3	0	0	(3)		2	1	1	(4)					
Groningen U.	1	1	0	1	1 + 1	1	1	0	1	2 + 0	(1)		50	
Leiden U.	2	0	1	5	(5)	4	0	0	6	(8)				
Utrecht U.	2	0	0	3	1 +	2	1	0	4	1 +		70	100	
Scandinavia	11	7	0	11	(7)	15	11	0	17	10				
Copenhagen Polyt.	1	0	0	(3)	(3)	1	0	0	4	(4)				
Copenhagen U.	1	0	0	(0)	(0)	1	1	0	0	(0)	0			
Oslo U.	2	1	0	2	(1)	3	1	0	4	(1)	0	15	9	
Lund U.	2	2	0	3	(1)	2	2	0	3	(1)	1	15	24	
Stockholm Högsk.	2	1	0	0	(0)	3	2	0	0	(1)				
Uppsala U.	2	1	0	2	(1)	3	3	0	3	(2)	1	24	45	
Helsinki U.	1	2	0	1	(1)	2	2	0	3	(1)			(30)	
Switzerland	11	2	4	12	10	12	4	7	16	15				
Basel U.	2	0	1	1	0 + 1	2	1	1	1	1 + 1				90
Bern U.	1	0	2	1	1	1	1	1	1	1				
Fribourg U.	1	0	0	2	0 + 1	2	0	0	4	2 + 1	4		14	
Geneva U.	1	2	0	2	2	1	0	1	3	2	(3)	47	69	
Lausanne U.	1	0	0	1	(½)	1	1	0	1	(1)	2		24	
Neuchâtel U.	1	0	0	1	(½)	1	0	0	1	(1)			(5)	(10)
Zürich ETH	3	0	1	3	(3)	3	0	2	3	3 + 1	7		150	
Zürich U.	1	0	0	1	1	1	1	2	2	1	1			

(Footnotes and sources on pages 25–28.)

TABLE A.4 (*continued*)

	Posts[a]										Students			
	1900					1910						In laboratory courses		
	Professor[b]	Jr. faculty[b]	Privatdoz.[b]	Assistant[c]	Employee[d]	Professor[b]	Jr. faculty[b]	Privatdoz.[b]	Assistant[c]	Employee[d]	Advanced, 1900[e]	1890	1900	In Lecture Courses, 1900
United States	32	31	36	46		51	52	66	76		(150)			
California U.	1	1	3	3	2	2	3	(3)	(4)					
Chicago U.	2	0	3	1		2	3	0	1		27			
Clark U.	0	1	0	0	1	1	0	(0)	(0)					
Columbia U.	1	5	1	5	0	3	3	6	7		7			500
Cornell U.	1	3	4	5	2 + 2	2	4	7	10	2 +	10			
Harvard U.	3	1	0	3	1 + 2	4	2	(4)	6		10			360
Johns Hopkins	2	2	0	5		3	1	(0)	3		15	98	95	84
Illinois U.	1	2	1	0		1	4	2	3					
Mass. Inst. Tech.	1	2	3	5		4	3	5	5					
Michigan U.	1	3	2	5		3	4	5	4		6	250	656	
Minnesota U.	1	1	2	1	1	2	1	3	2	(3)	2	30	152	
Missouri U.	2	0	1	0		3	1	(0)	(2)					
Nebraska U.	1	2	1	(1)		3	2	1	6					
New York U.	2	(0)	(1)	(2)		1	2	(1)	(2)					
Northwestern U.	2	0	1	1		1	2	0	1		3	46	41	
Ohio State U.	1	1	2	(1)		1	3	2	(2)					
Pennsylvania U.	1	1	3	0	1 +	3	1	11	1		10			
Princeton U.	2	1	1	0		6	3	(5)	(2)	2 +		22	30	
Stanford U.	1	3	2	1		1	3	1	4		1		125	300
Wisconsin U.	2	1	2	(2)		2	2	6	9		7			
Yale U.	4	1	3	5		3	5	4	2		6			202

[a]Sources: *Minerva: Jahrb.*, supplemented by G.B., Board of Ed., *Reports*; the lists of physics courses and teachers in German-speaking universities in *Phys. Zs.;* Italy, Min. pubbl. istr., *Annuario;* France, Min. instr. publ., *Statistique, 4*(1900); France, Min. educ. nat., personal communication (13 Apr. 1971); Weinberg, *Phys. dans 206 lab.* (1902); and by such catalogues, yearbooks, calendars, staff directories, etc., as are listed in the bibliography under the particular university or technical school. Here as elsewhere, blank entries signify no information; zeros, positive evidence of absence; parentheses, our guesses. We have generally excluded all institutes and teaching posts devoted solely to technical, applied, or industrial physics, but have included any men, e.g., the theoretical physicists signaled in note 26, occupying posts requiring concurrent attention to applications. We have used our judgment in deciding to list about one half of the men with double titles, like "Prof. of Physics and Geophysics"; we do not count professors emeriti.

[b]Professors and Junior Faculty:

Italy: "Incaricati" (men in part-time positions who usually held another full-time post) are counted as one half.

Scandinavia: "Junior faculty" includes the docent, lektor, laborator; the amanuensis is considered a (teaching) assistant.

Privatdozenten: For the United Kingdom and the United States we use this column, improperly, to list, respectively, Fellows and instructors. Our numbers of Fellows at Oxford and Cambridge in 1910 we derive from the first *Commonwealth Univ. Yearb.* (1914). For 1900 no reliable source has come to light; the figures are our guesses.

[c]Assistants. In general we rely heavily on Weinberg (1902).

Austria: Vienna U.: plus one Adjunkt and one assistant in the Radium Inst. in 1910. Kalinka (1911), p. 79.

British Empire: We count all assistant lecturers and demonstrators salaried at less than £125 (2500 marks) per annum as assistants; those above this level we consider to be junior faculty. For Oxford: Chapman, "Notes" (1973).

France: We include here the chef des travaux, of which there was one at every university except Clermont and Grenoble, which had none, and Nancy, which had two.

Germany: Berlin T.H.: 7 assistants in 1895. Prokop, *Ausbau* (1896), p. 29.

Scandinavia: Oslo U: of the assistants in 1910 two were V. Bjerknes' research assistants, paid by the Carnegie Institution, and two were K. Birkeland's research assistants, paid from his own pocket. Oslo U., personal communication (21 Mar. 1973).

[d]Employees. Includes janitors as well as technical assistants (mechanics, machinists, etc.). The latter are listed on the left, the former on the right, when their number is separately known. Men employed part-time or shared with other departments are counted fractionally.

Austria: Vienna U.: plus one mechanic and one servant in the Radium Inst. in 1910. Kalinka (1911), p. 79.

British Empire: Lodge, *Past Years* (1931), p. 147; Kay, *Nat. Phil., 1* (1963), 127–55; Cavendish Lab., *History* (1910), p. 82; Toronto, U., *President's Report* (1902), p. 32; Chapman, "Phys. at Oxford" (1973).

France: Paris, U., Conseil acad., *Rapports* (1906–07), pp. 101–03; (1910–11), p. 149; E. Curie, *Mme. Curie* (1937), p. 173; P. Langevin, *Revue du mois, 2* (1906), 12; Rocard in *Abraham. Commemoration* (1969), p. 11; Montpellier, U., *Livret* (1902–03), pp. 122–24; du Bourget, "Marseille" (1900), p. xv.

Germany: Glasser, *Röntgen* (1934), p. 78; Röntgen, *Briefe an*

Notes to TABLE A.4 (*continued*)

Zehnder (1935), p. 118; Leipzig, U., *Festschrift, 4,* part 2 (1909), 57; Lenz, *Gesch. U. Berlin, 3* (1910), 292, 296; Alsace-Lorraine, *Landeshaushaltsetat* (1900), pp. 108–09; Bavaria, Landtag, *Verhandl.* (1899–1900), Beilagenband IV, Budget 26, pp. 312, 315; Lehmann, *Phys. Inst. Karlsruhe* (1911), pp. 84–87; Württemberg, Kammer der Abgeord., *Verhandl.,* 34. Landtag (1899), Beilagenband 1, Heft VI, pp. 92–95, 238–39; Marburg, U., *1527–1927*, pp. 761–63; Darmstadt, T.H., *Gebäude* (1895), p. 90; Königsberg, U., *ZBBV,* 7 (1887), 13.

Italy: Min. pubbl. istr., *Annuario* (1900).

Scandinavia: Copenhagen, U., *Årbog* (1908–09), p. 383; Copenhagen, Polytekn. Laereanstalt, *Samlinger* (1910), p. 13 (note that the university had no laboratory; its students used the laboratory at the T.H.); Oslo, U., *1911–1961, 1* (1961), 510; Oslo, U., personal communication (21 Mar. 1973); Lund, U., *Årsberattelse* (1909–10); Leide, *Fysiska inst.* (1968), pp. 133–35; Uppsala, U., *1872–1897, 2* (1897) 153–54; Sweden, Utbildningsmin., personal communication (27 Mar. 1973); Helsinki, U., *Redogörlse* (1908–11), p. 68.

United States: Webster in *Clark U., 1889–1899,* p. 95; Howe and Grantham, "Physics at Cornell," pp. 19–20; Erikson, "U. of Minnesota Dept. of Phys.," p. 266; Birge, "Physics Dept., Berkeley," *1* (1966), v, 8, 14; Morison, ed., *Harvard U. 1869–1929*, pp. 284–6; McClenahan, "Princeton," *Science, 32* (1910), 293.

[e]Advanced Students. Students probably expected to do research.

Austria: Küchler (1906), p. 192, found 2 at U. Graz and 12 at U. Vienna in 1904.

British Empire: Aberdeen: $\Delta = 0$ and 6 math. and nat. phil. honors degrees with $\Delta = 0$, from *Calendar* (1902–03, 1909–10). Bristol: number of students in Physics Dept. static, from *Meeting of Governors* (1899–1901). Cavendish: about 12 research students estimated for 1900 from *History* (1910), pp. 221, 324–34. Edinburgh in 1903: G.B., Parl., H. of C., *Sess. Papers* 1903, *53*, 519. London, Royal Coll. Sci.: one research and/or fourth-year student in 1907–08, 3 in 1908–09, in a new building intended to accommodate 40–50 fourth-year, and 20 research students, from G.B., Parl., H. of C., *Sess. Papers* 1906, *31*, 565–66; 1909, *19*, 16. Manchester: plus 25 physics students, estimated for 1900 from Manchester, U., *Phys. Lab.* (1906).

France: In Paris in 1910 the number of affiliated researchers other than teaching personnel: U. Paris, 50 (20 in Lippmann's lab, 18 in M. Curie's, 12 in Bouty's); École normale, 5 (Abraham's lab); PCN, Paris, zero (Janet's and Sagnac's labs); France, Min. instr. publ., *Enquêtes*, Nr. 101 (1911), pp. 25–27.

Germany: "Doktoranden" and "selbständig Arbeitende": Berlin, U.: 30 in 1904. Küchler (1906), p. 192. Bonn: $\Delta = 3$/year, after 1904 limited to 13 by lack of space. *Chronik* (1900–01), pp. 42–43; (1904), p. 67; (1910), p. 50. Breslau: 15 in 1911. Lummer in *Festschr., 2* (1911), 447. Freiburg i. Br.: Küchler (1906), p. 193. Giessen: 1902–03. Lorey, *Nachr. Giessener Hochschulges., 15* (1941), 126. Halle: *Chronik* (1901–02), pp. 43–44. Leipzig: $\Delta = 1.5$/year, reaching 18 in 1908. *Festschr., 4*, pt. 2 (1909), 57. Marburg: c. 12 in 1910. *Chronik.* Munich, U.: 7 in 1893, 14 in 1904. Bayerisches Hauptstaatsarchiv, MK 11316. Kiel 1894–98: average of 5–6 students in the Fortgeschrittenenpraktikum. Schmidt-Schönbeck, *Kiel* (1965), pp. 103–04.

Italy: "inscritti per laurea in fisica," 1900. Ferraris, *Annali di statistica*, ser. 5, no. 6 (1913), x, 62, 87.

Netherlands: Groningen: 4 doctoral dissertations 1901–10. *Acad. Groningana, 1614–1914*, p. 492.

Scandinavia: U. Copenhagen and U. Oslo each produced two

doctoral students in physics between 1890 and 1910. Copenhagen, Polytekniske Laereanstalt, *Samlinger* (1910), p. 10; Oslo U., *1911–1961, 2* (1961), 300–03. Data for Swedish schools from Beckman and Ohlin, *Forskning* (1965), p. 39.

Switzerland: Fribourg: doubling to c. 8 in ten years. *Bericht* (1899–1900), p. 26; (1910–11), pp. 21–26; Fribourg (Canton), Dir. de l'instr. publ., *Compte rendu* (1900), p. 145. Geneva: *Historique* (1914), p. 22. Lausanne: Vaud (Canton), Dept. de l'instr. publ., *Compte rendu* (1900), p. 30. Zürich: 8 doctorates conferred by the University of which 7 upon students at the ETH. Zürich, U., personal communication (19 Apr. 1973).

United States: Except where another source is given, figures are numbers of "graduate students" in 1903 from U.S., Office of Ed., *Report, 2* (1903–04), 1424. Harvard: graduate students in 1899, Δ = 0, from *President's Report* (1899/1900). Michigan: 6 graduate students in 1899–1900, 3 in 1904–05, from *Catalog* and *President's Reports*. Minnesota: 5 graduate students in 1894–95, 2 in 1899–1900, 3 in 1904–05, 3 in 1909–10, from *Catalogs* and *President's Reports.* Northwestern: 3 graduate students in 1899–1900, 2 in 1909–10, from *Catalogs* and *President's Reports*. Pennsylvania: *Provost's Report.*

[f]Students in laboratory courses. Unless otherwise indicated all entries are from Weinberg, *Phys. dans 206 lab.* (1902).

British Empire: Edinburgh, Aberdeen in 1903: G.B., Parl., H. of C., *Sess. Papers* 1903, *53*, 519. Aberdeen: doubling by 1907–08. *Calendar* (1909–10). Manchester: Δ = 5 per year. *Calendar* (1900–01, 1909–10). Nottingham, *Calendar* (1897–98), p. 65, implies still only c. 50. Oxford: Chapman (1973). McGill: *Annual Report* (1897), p. 26; (1902–03), p. 35. Toronto: *President's Report* (1902), p. 13.

France: Paris, U.: Δ = 10/year. Conseil de l'U., Conseil acad., *Rapports* (1906–07), pp. 101–05. The division of the students between the faculty of science and the PCN courses was: Besançon, 12 + 25; Caen, 15 + 20; Dijon, ? + 67; Grenoble, 10 + 35; Lille, 16 + 70; Marseille, 14 + 60; U. de Paris, 100 + 160; Toulouse, 7 + 120.

Germany: Darmstadt: Cf. Lexis, *Unterrichtswesen, 4*, pt. 1 (1904), 283–84, who gives 292 students in the physics Praktikum (four afternoons per week) in 1902–03. Munich, T.H.: rise 1890–1900 followed by decline (Δ = – 4/year). *Bericht* (1900–01), p. 16; (1908–09), p. 24. Berlin, U.: Δ = 10/year; advanced course c. 15, Δ = 0; course for pharmacists, Δ negative. *Chronik* (1900–01), p. 130; (1909–10), p. 174; (1910–11), p. 200. Bonn: *Chronik* (1900–01), pp. 42–43, and (1910), p. 50, confirms Weinberg and yields Δ = 5/year. Breslau: *Chronik* (1900–01, 1910–11) gives only half of Weinberg's 65 and yields Δ = 10/year. Halle: Δ = 10/year. *Chronik* (1901–02), pp. 43–44; (1905–06), p. 61. Kiel: Δ = 7/year. *Chronik* (1899–1900), p. 46; p. 68. Königsberg: *Chronik* (1899–1900), p. 33. Leipzig: *Festschr, 4,* pt. 2 (1909), 57, reports 116 versus Weinberg's 155. Marburg: Δ = 15/year. *Chronik* (1900–01), p. 37; (1903–04), p. 33. Rostock: *Verzeichnis der Behörden* (1900–01), p. 29.

Scandinavia: Leide, *Fysiska institutionen*, p. 134; Lund, U., *Årsberattelse* (1900); Uppsala, U., *Redogörlse* (1900), *Handlingar* (1903), p. 20, and *Uppsala universitet, 1* (1897), 516–17; Norway, Kirke- og undervisningsdepartementet, *Universitets- og skoleannaler* (1900–01); Helsinki, U., *Redogörlse* (1905–08).

Switzerland: Fribourg: Δ = 1/year. *Bericht* (1900–01), p. 21; (1910–11), pp. 21–22. Neuchâtel (canton), *Rapport annuel: enseignement sup.* (1910–11).

[g]Students in lecture courses (estimated in most cases from the number of seats in the main lecture hall of the physical institute).

British Empire: Aberdeen, Edinburgh: students in nat. phil. 1903

Notes to TABLE A.4 (*continued*)

from G.B., Parl., H. of C., *Sess. Papers* 1903, *53*, 458–59, 519. Bangor: *Calendar* (1900–01). Belfast: $\Delta = 7$/year. *Pres. Report* (1900–01, 1908–09). Bristol: rising rapidly. *Meeting of Governors* (1899–1901). Cork: *Pres. Report* (1899–1900). Dublin: *Calendar* (1911–12), p. 326. Galway: *Pres. Report* (1899–1900), p. 12. Leeds: *Ann. Report* (1899–1900), p. 43. London, R. Coll. Sci.: 94 individual students in physics, of which 83 in the first year's course. G.B., Bd. of Educ., *Report on the Museums* (1901), pp. 54–55. Manchester: *Calendar* (1900–01), p. 469. Toronto: *Ann. Report* (1901–02), p. 13.

France: Lille: Moisson and Matignon, *RGS, 6* (1895), 477–93. Montpellier: *Livret* (1902–03), p. 123; (1907–08), p. 214. Paris, École Polytech.: Callot, *Hist.* (1958), p. 115.

Germany: Karlsruhe: Lehmann, *Phys. Inst. Karlsruhe* (1911) p. 78 (giving detailed time series showing *decline* to half this maximum by 1911). Stuttgart: Württemberg, Kammer der Abgeord., *Verhl.,* 37. Landtag (1907), Beilagenband 1, Heft XV, pp. 17–18. Berlin, T.H.: Küchler (1906), p. 198. Berlin, U.: 270 in exp. lecture, $\Delta = 7$/year; 90 in theoretical lecture, $\Delta = 10$/year. *Chronik* (1900–01), pp. 130–33; (1910–11), pp. 200–03. Breslau: Lummer, in Breslau, U., *Festschr., 2* (1911), 447. Darmstadt: Lexis, *Unterrichtswesen, 4,* part 1 (1904), 283–84. Erlangen: Wiedemann, *Phys. Inst. U. Erlangen* (1896), p. 14. Giessen: Wien, *Phys. Zs., 1* (1899), 155–60; Küchler (1906), p. 198. Kiel: *Chronik* (1899–1900), p. 46. Leipzig: *Festschrift, 4,* pt. 2 (1909), 57, 68. Marburg: $\Delta = 10$/year. *Chronik* (1907–08), p. 35. Tübingen: Paschen to Naturwiss. Fak., 20 Feb. 1906 (U. Archiv, Tübingen, 117/904).

Switzerland: Basel, U., *1887–1913*, pp. 116–18. Neuchâtel: see note f.

United States: Columbia: *Report of the President* (1901–02); 465 individual students in physics courses in 1897, $\Delta = 25$/yr., from *Report of the President, 8* (1897), 213. Harvard: 360 students in undergrad. physics courses in 1899–1900, $\Delta = 0$, from *President's Report* (1899–1900). Johns Hopkins: *Annual Report* (1902–03) gives time series of attendance upon physics courses showing steady downward trend, $\Delta = -6$/yr. Stanford: *Annual Register* (1904–05), p. 45. Yale: 202 "sophomore options" in 1900–01, with $\Delta = 0$, from *President's Report* (1903–04), p. 58.

come a requirement for agronomists, engineers, physicians, and teachers of science in secondary schools.[18] It appears that intending secondary school teachers were the most important factor in the growth rate in the 1880's, and that engineers and, especially, medical students had overtaken them by the turn of the century.[19] But although the number of students in elementary physics courses increased nearly monotonically, that was by no means true of those concentrating on physics. The number of science students in German universities decreased in the late eighties and early nineties owing to a "gloomy outlook for the future of natural scientists and mathematicians" particularly in secondary school teaching.[20] The number of physics students at the Italian universities declined after the turn of the century, probably because industrial employers preferred graduates of technical schools.[21] At the same time in the United States even

[18]Cf., Cardwell, *Sci. in England* (1957), pp. 100, 167, 182; Sanderson, *Univs. and British Industry* (1972), pp. 43, 91, 101; Beier, *Die höheren Schulen* (1909), pp. 517-29; Conrad, *German Univs.* (1885), p. 163; Paris, U., Conseil acad., *Rapports* (1902-03), pp. 98-99, 104, and *ibid.* (1911-12), pp. 181, 211; Toronto, U., *President's Report* (1909), p. 6. The laboratory arrangements for this clientele are discussed by Cajori, *Hist. of Phys.* (1962), pp. 387-406; Cavendish Lab., *History* (1910); Weinberg, *Phys. dans 206 lab.* (1902); and infra, Section D.

[19]Appell, *Revue de Paris, 17,* No. 6 (1910), 99; Küchler, "Phys. Labs. in Germany" (1906), pp. 188-89; Lexis, *Unterrichtswesen, 1* (1904), 253; Wiedemann, *Phys. Inst. U. Erlangen* (1896), p. 7. Some scattered data will suggest the numerical importance of the medical students in the elementary physics laboratories: in 1885 the Cavendish had 90 students, in 1888 153, the increase owing entirely to opening of courses for medical students. Fitzgerald in Cavendish Lab., *History* (1910), p. 9. At Columbia in 1901/2 some 500 students worked in the labs, many coming from the College of Physicians and Surgeons, which had recently discontinued its own physics courses. Columbia U., *Report of President* (1896-97), p. 5; *Columbia U. Quarterly, 4* (1901-02), 319. At Leipzig in 1908 72 of 188 students in the elementary physics labs intended to make careers in medicine or pharmacy. Wiener in Leipzig, U., *Festschrift, 4,* Pt. 2 (1909), 57. For the French P.C.N. system, infra.

[20]Röntgen to Hertz, 27 September 1888, in Glasser, *Röntgen* (1934), pp. 97-99; to Zehnder, 23 April 1889, in *Briefe an Zehnder* (1935), p. 9; in the latter year (1889) there was a spate of pamphlets on "Die Überfüllung der gelehrten Fächer." Würzburg, U., Inst. f. Hochschulkunde, *Bestände der Bibliothek* (1968), E. 50-60. Lorey, *Studium der Math.* (1916), p. 22, shows that the number of students enrolled in the period 1880-1914 in mathematics and natural sciences at U. Göttingen closely paralleled the number of candidates passing the Prussian state examinations for teachers in those fields. Both curves have deep troughs in the late 1880's and early 1890's.

[21]Ferraris, *Riforma sociale, 13* (1903), 896; *ibid., 17* (1907), 738; cf. the rector's address in Rome, U., *Annuario* (1898-99), p. 6. The effect of industry in suppressing the growth of academic physics in Italy is also emphasized by Corbino, Soc. ital. prog. sci., *Atti, 5* (1911), 304, and Enriques, *Scientia, 3* (1908), 134. The same effect occurred in Sweden; Linders and Lander, *Statistisk undersökning* (1936), p. 21.

as the number of doctorates rose the number of students was apparently falling at all levels, graduate, undergraduate, and secondary school.[22]

The posts we have counted divide (although not so cleanly as we might like) between experimentalists and an inhomogeneous group known as theoretical or mathematical physicists. The theoretical physicists properly so-called existed primarily in German speaking countries and in those, like Sweden and the Netherlands, whose academic ranks followed the German model. The incumbents of the few French chairs of physique mathématique, like H. Poincaré or J. Boussinesq, tended more toward mathematics than toward physics. In the United States mathematical (or theoretical) physics was very weak in 1900. Tiny Clark University could well take as its "principal claim" to distinction its stress upon the subject, "which has not yet become popular in this country."[23] Britain is a special case to which we shall turn momentarily.[24]

Table A.5 gives a measure of the vigor of theoretical physics in the several nations. One should remember that theoretical physics emerged as a

[22]F. Sanford in Stanford U., *Report of the President* (1906-07), pp. 41-42, attributed the paucity of advanced students to the demand for secondary school teachers. There are, however, various indications of at least a relative dip in the study of physics in secondary schools in the preceding fifteen years. Hahn, *Zs. phys. u. chem. Unterricht, 20* (1907), 189.

[23]Webster in *Clark U., 1889-1899,* p. 91. Coben, *Am. Hist. Rev., 76* (1971), notwithstanding, the Americans, well aware of their deficiencies in theoretical physics some twenty years before the advent of quantum mechanics, had already begun concerted efforts to foster interest in the subject and import its practitioners in the first decade of the century. Cf. E.F. Nichols to A. Sommerfeld, 20 July 1905 (SHQP, mf. 32); Milne, *Jeans* (1952), pp. 12-14; the career of Jakob Kunz; and the series of lecture courses in mathematical physics at Columbia University from 1905 onward by Bjerknes, Lorentz, Planck, Wien, Runge, i.a., published by the Ernest Kempton Adams Fund for Physical Research. In many American universities (e.g., Stanford, Cornell) instruction in mathematical physics was given in the mathematics (or applied mathematics) department. The situation at Columbia was more promising: a few advanced students read theoretical physics in a special seminar which, it was hoped, would become a regular course in the physics department. *Columbia U. Quarterly, 4* (1901-02), 319. The directors and veterans of such seminars no doubt made up a substantial portion of the men in ordinary physics posts identified by Cattell, *Amer. Men of Sci.* (1906), as working in theoretical physics.

[24]For distinction and distance between experimentalists and theoreticians see Rayleigh, *Rayleigh* (1968), pp. 46, 132, 317-18; Ramsay, *Nature, 65* (1902), 510-11; Lodge, *Past Years* (1931), 151-52; Lamb, B.A., *Reports* (1904), pp. 421-31; Tannery, *RIE, 22* (1891), 49-95 and *35* (1898), 46; Darboux, *RIE, 35* (1898), 263-67; Voigt, *Phys. Forschung* (1912), pp. 12-13; Kevles, *Phys. in America* (1964), pp. 135-38, 243-44.

TABLE A.5

Posts in Theoretical and Mathematical Physics[a]

	1900				1910			Adressbuch[c] (1909)
	Senior faculty	Junior faculty[b]	Total	Per-cent[d]	Senior faculty	Junior faculty	Total	
Austria-Hungary	4	4	8	17	7	2	9	13
T.H.	0	0	0	0	0	0	0	
Univ.	4	4	8	22	7	2	9	
Belgium	1	0	1	10	0	2	2	2
British Empire	0	2	2	3	0	2	2	2
U.K.	0	2	2	3	0	2	2	2
Other	0	0	0	0	0	0	0	0
France	4	0	4	8	4	0	4	2
Germany	8	8	16	15	10	6	16	16
T.H.	3	1	4	15	4	0	4	
Univ.	5	7	12	16	6	6	12	
Italy	4	4	8	19	6	6	12	11
Japan	0	0	0	0	2	1	3	0
Netherlands	1	2	3	25	2	1	3	2
Russia	1	0	1	3	1	0	1	2
Scandinavia	3	1	4	20	5	2	7	4
Switzerland	2	1	3	17	2	1	3	4
United States	2	1	3	3	0	1	1	2

[a]Same sources as Table A.4.
[b]Including Privatdozenten.
[c]*Adressbuch der lebenden Physiker* (1909).
[d]As percent of all physics posts.

separate discipline only in the last third of the nineteenth century and that, in Germany in particular, it had to overcome considerable institutional obstacles, like unfriendly professors of experimental physics.[25] Instruction in theoretical physics was generally regarded as supplemental to that in experimental physics; consequently it was common to establish

[25]Schmidt-Schönbeck, *Phys. Kieler U.* (1965), p. 100; Hartmann, *Planck* (1953), pp. 14–15. Cf. the case of Janne Rydberg, preferred by the faculty of U. Lund for their vacant professorship of physics over Victor Bäcklund, his senior and the incumbent extraordinary professor for mathematical physics, on the ground that Bäck-

theoretical physics at an institution by setting up an Extraordinariat, not a full professorship. Often these positions went to experimentalists who merely taught theoretical physics;[26] hence the nominal strength of the theoreticians substantially exceeded their actual strength—especially since, if the experimentalist occupying the theoretical Extraordinariat did not rise, he might tie it up for life. When the theoretician did obtain a chair he often enjoyed the perquisites of his experimentalist colleagues, including an "institute" (usually in buildings shared with experimentalists), full-time assistants, and perhaps laboratories and a workshop.[27]

The British case is unusual because about half of the physics chairs in England, Scotland, and Wales were held by former wranglers, men who had obtained first-class degrees in mathematics or, better, in applied mathematics at Cambridge.[28] These men, who gave British physics its distinctive character,[29] maintained their dominance until after 1910, as appears from Table A.6(1). A similar hegemony was exercised in France by normaliens and polytechniciens: as one sees from Table A.6(2), the École normale, like Cambridge, furnished about half of all academic physicists in its country, and the École polytechnique made up much of the balance.[30] The

---

lund knew too little experimental physics for the chair; Bäcklund complained to the government, which gave him the professorship. Leide, *Kosmos, 32* (1954), 15-32, and *Fysiska institutionen* (1968), pp. 129-30; Weibull, *Historia, 4* (1968), 350-51. Since a high proportion of young theoretical physicists were Jewish they also had to contend with academic anti-semitism. Cf. Conrad, *Universitätsstudium* (1884), p. 138; Röntgen, *Briefe an Zehnder* (1935), pp. 29-30; Busch, *Gesch. Privatdoz.* (1959), pp. 155-57, 162; Preston, *Science and German Jews* (1972), pp. 196-215; Forman, *Phys. in Weimar Germany* (1968), pp. 114-21; Forman, *Minerva, 12* (1974), 56.

[26] E.g., Mathias Cantor (Würzburg), Th. Des Coudres (Leipzig), C. Fromme (Giessen), W. Kaufmann (Bonn), J. Koenigsberger (Freiburg). It was moreover not uncommon for the theorist, or his post per se, also to carry responsibility for instruction in *applied* physics, as in the cases of H.F. Weber (ETH), Des Coudres (Göttingen), H. du Bois (Utrecht). Cf. the remarks of E. Fues, a theoretical physicist, on the "separation of this discipline which until recently had been united with applied physics . . . in many *Hochschulen.*" Hannover, T.H., *Festschr.* (1931), p. 42.

[27] Des Coudres in Leipzig, U., *Festschrift, 4,* Pt. 2 (1909), 60-69; Voigt in Göttingen, U., *Phys. Inst.* (1906), p. 38; Planck in Lenz, *Gesch. U. Berlin, 3* (1910), 276-78; Forman and Hermann, art. "Sommerfeld," *DSB;* Prague, U., *Deutsche Univ. Prag* (1899), pp. 405-06; and, for T.H. Vienna, Austria, Min. f. Kultus, *Neubauten* (1913), p. 46.

[28] For the wrangling system see Glaisher in London Math. Soc., *Proc., 18* (1886), 4-38; Rouse Ball, *Hist. Math. at Cambridge* (1889), pp. 187-219; Roth, *American Math. Monthly, 78* (1971), 223-36.

[29] We have in mind the school of Kelvin, Stokes, and Maxwell.

[30] Paris, École normale supérieure, *Centenaire* (1895); Paris, École polytechnique, *Centenaire* (1894/7); Peyrefitte, ed., *Rue d'Ulm* (1963).

TABLE A.6

(1) Ratio of Chairs Held by Wranglers to Total Physics Chairs[a]

	1850	1860	1870	1880	1890	1900	1910	1920
England	3/7	5/10	4/12	6/15	10/22	12/24	12/24	9/32
Scotland	3/5	3/4	4/5	4/5	4/5	2/5	1/6	1/7
Ireland	0/3	0/5	0/7	1/9	1/9	3/9	3/9	3/8
Wales	0/0	0/0	0/0	0/1	0/3	1/3	1/4	0/5
TOTAL	6/15	8/19	8/24	11/30	15/39	18/41	17/43	13/52
Percent Wrangler	40	42	33	37	37	44	39	25

(2) Percentage of French Physicists c. 1900 Educated at the Grandes Écoles[a]

	École normale	École polytech.	Other	Unknown	Sample size
Major Paris chairs	52	31	17	0	28
All faculty	49	14	18	19	94

[a]From samples described in notes to Table B.2.

proportions of men from these schools were larger in the major centers (Cambridge, Paris) than elsewhere. The dominance of Paris appears even stronger on the next higher level, that of the Dr.-ès-sciences of the Sorbonne, a degree held by nearly all French academic physicists.

No such center of power existed in other countries. Berlin, perhaps the leading German school, turned out about one new Privatdozent in physics every year around 1900, as compared with six Sorbonne physics doctorates.[31] Undergraduates in Germany were highly migratory; 70 percent of a sample of ninety-three German physicists attended at least two universities as undergraduates, 30 percent attended three or more.[32] As for the United States, of sixty-four men who entered the physics profession between 1895 and 1906, fifteen came from Johns Hopkins, eleven from Cornell, and seventeen from Harvard, Columbia, and Chicago together.[33]

[31] See Table A.3.

[32] For our sample see infra, Table B.2.

[33] Kevles, *Phys. in America* (1964), pp. 309–12. In the period 1906–16 the institutional origin of American Ph.D.'s was more diverse.

TABLE A.7

Distribution of Physicists by Nation and Field, 1909[a]

	Physics			Physics and mathematics			Physics and other			Mathematical physics	Theoretical physics
	Secondary schools	Colleges and universities	Neither	Secondary schools	Colleges and universities	Neither	Secondary schools	Colleges and universities	Neither		
Austria-Hungary	3	54	3	7	4	2	1	5	0	9	3
Austria only	2	42	2	7	2	0	1	2	0	8	3
Hungary etc.	1	12	1	0	2	2	0	3	0	1	0
Belgium & Luxemburg	0	6	5	0	0	1	0	0	0	2	0
British Empire	1	154	15	0	15	1	0	8	0	2	0
U.K.	1	110	15	0	4	2	0	5	0	2	0
India	0	24	0	0	5	0	0	3	0	0	0
Other	0	20	0	0	6	0	0	0	0	0	0
Bulgaria	0	1	0	0	0	0	0	0	0	0	0
China	0	0	0	0	0	0	0	0	0	0	0
France	20	124	6	0	0	0	0	7	1	1	0
Germany	74	195	90	178	3	5	3	5	4	6	10
Greece	1	1	0	0	0	0	0	1	0	0	0
Italy	25	50	13	0	4	0	0	3	0	11	0
Japan	0	10	0	0	0	0	0	0	0	0	0
Latin America	0	8	0	0	1	0	0	3	0	0	1
Netherlands	1	16	4	1	0	0	0	0	1	2	0
Rumania	0	4	0	0	1	0	0	0	0	0	0
Russia	2	46	5	2	1	0	0	4	0	2	0
Scandinavia	1	20	4	1	0	0	0	1	0	4	0
Denmark	0	7	1	0	0	0	0	0	0	0	0
Finland	0	0	1	0	0	0	0	0	0	0	0
Norway	0	3	1	0	0	0	0	0	0	1	0
Sweden	1	10	1	1	0	0	0	1	0	3	0
Spain & Portugal	0	19	0	0	1	0	0	2	0	1	1
Switzerland	0	23	6	3	0	1	0	3	0	2	2
Turkey	0	0	0	0	0	0	0	0	0	0	0
United States	4	155	8	0	9	0	0	4	0	2	0
TOTAL	132	886	159	192	39	11	4	46	6	44	17

[a]From *Adressbuch der lebenden Physiker* (1909).

TABLE A.7 (*continued*)

Mechanics	Applied mathematics	Geophysics	Geodesy	Meteorology	Astronomy	Astrophysics	Biophysics	Medical physics	Electrochemistry	Technical and industrial physics	Electrotechnology	Other engineering	Photography and optics	Physical chemistry	Other[b]	TOTAL
9	0	3	12	33	36	8	0	1	3	2	19	0	0	17	9	243
1	0	3	8	20	28	4	0	0	2	1	15	0	0	14	7	172
8	0	0	4	13	8	4	0	1	1	1	4	0	0	3	2	71
5	0	0	1	2	14	0	0	0	1	2	4	0	0	4	3	50
15	5	0	6	15	91	4	0	0	1	0	29	18	1	18	7	408
12	4	0	2	9	53	4	0	0	1	0	22	13	1	17	5	282
0	0	0	0	5	13	0	0	0	0	0	1	2	0	0	1	54
3	1	0	4	1	25	0	0	0	0	0	6	3	0	1	2	72
0	0	0	0	1	1	0	0	0	0	0	0	0	0	1	0	4
0	0	0	0	2	3	0	0	0	0	0	0	0	0	0	1	6
14	5	0	0	25	52	2	1	5	3	4	18	11	1	6	10	316
18	2	8	26	35	90	13	0	2	17	3	49	11	13	75	28	962
0	0	0	1	0	1	0	0	0	0	0	0	0	0	1	0	6
22	0	4	12	15	37	3	0	1	2	3	9	0	0	4	3	221
0	1	2	0	1	2	0	0	0	0	0	6	0	0	0	0	22
5	0	0	5	4	18	0	0	2	0	1	3	2	0	3	3	59
2	0	0	0	6	10	0	0	1	2	0	2	0	0	6	2	56
0	0	0	0	2	2	0	0	0	0	0	0	0	0	1	0	10
20	2	0	10	19	42	6	0	1	1	0	8	0	0	14	10	195
1	1	0	2	11	13	0	0	0	3	0	3	0	0	6	3	74
0	0	0	1	4	3	0	0	0	1	0	3	0	0	0	1	21
0	0	0	0	0	0	0	0	0	0	0	0	0	0	0	0	1
0	1	0	0	2	1	0	0	0	1	0	0	0	0	0	2	12
1	0	0	1	5	9	0	0	0	1	0	0	0	0	6	0	40
3	0	1	2	6	15	2	0	0	0	0	0	0	0	6	0	59
2	1	1	0	1	10	0	0	0	1	0	6	0	0	7	4	73
0	0	0	0	0	2	0	0	0	0	0	0	0	0	0	0	2
18	9	0	2	9	111	4	0	1	0	0	32	11	0	11	14	404
134	26	19	79	186	550	42	1	14	34	15	188	53	15	180	98	3170

[b] Aeronautics, acoustics, calibration, geodetic surveys, hydrography, mineralogy, navigation, physical geography, physiology, seismology, and weights and measures.

2. Contributors to Basic Physics

It is plain that our count of physics posts falls short of the number of men engaged in pure research in physics in several ways. For one, we have ignored university posts explicitly reserved for other faculties whose incumbents, however, might well publish much of their work in physics journals. For example, we have not counted the mathematics chairs held by Volterra, Levi-Civita, or Larmor or by former wranglers awaiting translation to physics posts.[34] We have systematically ignored the chairs devoted to mechanics whose incumbents generally, but by no means exclusively, attended to mathematics in France and to technology in Germany. Similarly we have left aside all posts in electrotechnology, in physical chemistry,[35] and in atmospheric physics, geophysics, and meteorology. Nor have we counted physics posts in medical schools, although we have admitted instructors in the courses for medical students given within science faculties, such as the French P.C.N. courses (*sciences physiques, chimiques et naturelles*).[36]

Apart from the higher schools a variety of scientific institutions sheltered contributors of research papers: observatories, weather bureaus, standards laboratories,[37] geodetic surveys, seismological stations, and technological institutes. Other contributors included secondary school teachers like Elster and Geitel, men of independent means with private laboratories,[38] and the occasional engineer or industrial physicist with the time

[34]We find that six out of 90 British physicists active c. 1900 held a post in mathematics at some point in their careers. For definition of the sample see notes to Table B.2.

[35]Consequently we omit men like Nernst and Perrin. As these names suggest, physical chemical posts might shelter important physicists. We accordingly give the number of professorships/junior faculty in the subject in 1900 as compiled from *Minerva: Jahrbuch:* Austria, 1/1; Belgium, 1/0; France, 1/1; Germany 1/6 (plus two professors in T.H.'s); Italy, 1/1; Netherlands, 0/1; Russia, 0/2; Sweden, 0/1; Switzerland, 0/2. Cf. however the penultimate column of Table A.7.

[36]The P.C.N. courses were established in 1893. Liard, *Enseignement supérieur, 2* (1894), 396; France, Min. instr. publ., *Recueil des lois, 5* (1889-98), 379-84, 365. Some universities continued to maintain substantial labs for medical physics or biophysics, like that at Montpellier, which housed two faculty and two assistants in 1902. Montpellier, U., *Livret* (1902), pp. 109-10. We do not count such posts.

[37]Of these the Physikalisch-Technische Reichsanstalt was by far the largest contributor to basic physics. See Charlottenberg, PTR, *Wiss. Abhandl.* (1894+); *Bisherige Tätigkeit* (1904), etc. Cf. Voigt, *Phys. Forschung* (1912), and the ensuing exchange with Warburg, *Phys. Zs., 13* (1912), 1091-95.

[38]For the British amateur: Schuster, Brit. Ass. Adv. Sci., *Reports* (1892), pp. 627-35; Carhart, *Science, 12* (1900), 701; S.P. Thompson, Phys. Soc. London, *Proc., 17*

and taste for basic research. A measure of the relative research contributions of these men may be collected from Table A.8.

So much for errors of defect. One would also err by excess if one assumed that each post we have counted was held by a distinct contributor. The incumbent might, like the fainéant Professor of Experimental Physics at Oxford, R. B. Clifton, contribute nothing at all,[39] or cultivate something other than basic physics. One might devote oneself to the new specialties separating from physics at the end of the nineteenth century, like geophysics or biophysics, and, while occupying a post for general physics, publish in journals for geology or physiology. Again, some of our posts, at most 5 percent, were vacant in 1900. Finally one should correct (as we have done in Tables I and A.1) for multiplication of benefices, or the occupation of more than one post by the same individual. In France, where this multiplication, known as the *cumul,* was established practice, the ratio of distinct incumbents to counted posts was about 0.9 in 1900.[40]

We have not tried to count directly the number of contributors to basic physics at the turn of the century. As a maximum we offer, in Table A.7, a breakdown of the physicists listed in the *Adressbuch der lebenden Physiker* for 1909. The figures for academic physics agree quite well with our values for 1910 (or, what is the same thing, with the values for 1900 [Table A.1], increased by the amount to be expected from the growth

---

(1901), 12-25; G. B., Parl., H. of C., *Sess. Papers,* 1898 (Cd. 8977), *45,* 33, 44. The best known of the private laboratories were Rayleigh's and Crookes'. Rayleigh, *Rayleigh* (1968), pp. 149-50, 196-97; Fournier d'Albe, *Crookes* (1923), p. 303. The Royal Institution's Mond laboratory was constructed partly to support and encourage the "amateur" of all countries. Cohen, *Mond* (1956), pp. 219-21. For private laboratories in France: Etard, Soc. chim. Paris, *Bulletin, 31* (1904), i-viii (Demarçay); *RGS, 14* (1903), 1024 (Cornu); *DSB, 2,* 487-89 (M. de Broglie). For Germany: J.J. Thomson, *Recollections* (1937), p. 409; du Bois in *Elster-Geitel Festschrift* (1915), pp. 245-50; Lehmann, *Phys. Inst. Karlsruhe* (1911), p. 82; Crew, "Diary, 1895": "Nothing in [Viktor] Schumann's laboratory [at Leipzig] that is not of the very best make . . . very complete in every way . . . spent 50,000 marks on his apparatus. . . . This laboratory presents the finest combination of neatness, cleanliness, and practical working atmosphere I have *ever* seen or expect to see." And we should add Schumann's fellow amateur of vacuum spectroscopy, Hans Hauswaldt, a Magdeburg businessman, who lent Rowland gratings to the physicists at Göttingen, and whose widow in 1912 gave the University of Tübingen apparatus appraised at 8,500 marks. U. Archiv, Tübingen, 117/904; Gött., U., *Phys. Inst.* (1906), pp. 41, 65.

[39] Heilbron, *Moseley,* pp. 34–35.

[40] We find the *cumul* elsewhere, e.g., in J.J. Thomson's simultaneous tenure of professorships at Cambridge and at the Royal Institution. Laboratories were attached to both posts. Rayleigh, *J.J. Thomson* (1942), pp. 147-49.

TABLE A.8

Relative Contributions of Academic and Nonacademic Physicists circa 1900

	Domestic contributors[a]/contributions[b] (each as percent of total)						Percent foreign[c] contributors / contributions	Total number contributors[a] / contributions[b]
	Affiliated with:				Un-affiliated	Un-identified		
	Our "higher schools"	Other scientific institutions	Other "higher schools"	Secondary schools				
British Empire								
Phil. Mag.	68/68	4/10	12/9	2/1	12/10	3/2	15/12	223/425
France								
Journ. de phys.	61/70	6/7	5/4	7/9	6/4	13/7	25/20	149/277
Comptes rendus	62/65	8/10	3/3	5/5	9/6	13/12	12/9	99/149
Germany								
Annalen d. Phys.	71/77	13/11[d]	1/1	7/5	3/2	5/4	23/24[e]	130/175
Italy								
Nuovo cimento	67/71	3/2	11/14[f]	2/4	1/1	15/8	6/3	97/179
United States								
Phys. Rev.	59/66[g]	4/2	31/29	2/1	0/0	3/2	14/10	135/252

[a]Number of distinct individuals authoring or jointly authoring papers in *Phil. Mag.*, *45–50, 1–4* (1898–1902); *Journal de Physique*, *7–10, 1* (1898–1902); *Comptes rendu* (Paris), *130* (1900); *Annalen der Physik*, *2–4* (1900); *Physical Review*, *6–15* (1898–1902).

[b]Number of distinct papers published by these individuals in the journal and years in question.

[c]Foreign contributors include nationals working abroad; domestic contributors include affiliated foreign nationals. (Half the foreign contributors to the *Phil. Mag.* were Americans.) Unidentified contributors have been counted as "domestic."

[d]All but one of these contributors and contributions were affiliated with the PTR, Charlottenburg.

[e]12 of these 30 foreign contributors were in Austria-Hungary.

[f]7 of these 10 contributors, and 20 of these 24 contributions, were affiliated with 3 institutions: the Istituto di Studi Superiori, Florence, and the institutes of technology at Rome and Padua.

[g]In the case of the U.S. "our" institutions are the set of 21 leading universities. It will be noted that if this figure (59 percent) is multiplied by 3/2 one obtains very nearly the sum of these and the further contributors affiliated with other "higher schools" in the United States. This is a further justification of our "augmentation" of the numbers and funds of the 21 by one half to obtain figures for the U.S. as a whole.

rate for the decade 1900–1910 [Table A.2]).[41] These figures do not, however, give an adequate idea of the *relative* contribution of the academics. Table A.8 shows a more appropriate measure, namely the proportion of the papers published in the leading physics journals in the years 1898 through 1902 attributable to physicists of various affiliations. The great preponderance of contributors affiliated with our institutions justifies our decision to restrict ourselves to academic physics.[42]

[41]The only important discrepancy is for the United States, no doubt a consequence of disinterest in or ignorance of the *Adressbuch* on the part of younger American physicists.

[42]We note that the *Adressbuch's* figures for academic physics in the leading countries give c. 650 when projected back to 1900 by the rates of change of Table A.2. We get about the same number of post-holding physicists (Table A.1). Cf. Poggendorff, *4* (1904), which lists the works of 500 physicists. Other estimates: Watanabe, Congr. internat. hist. sci., 10th (1962), *Actes, 1,* 197–208 (Japan); Wind, Nederl. Natuur en Geneesk. Congr., *Handel.,* 7 (1899), 91–139, and subsequent years.

B. PERSONAL INCOME

1. Sources and Amounts

a. Salaries, Fees, and Fringes

We give in Table B.1 the ranges and averages of the salaries and fees attached to the teaching posts of Table A.4. Multiplying these averages by the number of posts gives the figures in Table I for the total amount spent in each country on salaries and fees for academic physics. As one sees, the *average* regular income (salary plus fees) attached to a chair was much the same in the various countries. Since, however, the cost of living in the United States was appreciably higher than in Europe, perhaps as much as 50 percent higher, the American professor (and, at the bottom of the ladder, the American assistant) fared less well than his European counterpart.[1] Among the Europeans the British enjoyed the highest salaries in all academic ranks, but especially in the lowest, and came into their positions soonest (Tables B.2 and B.3).

After the generally deflationary nineties, the Western world experienced an inflation of about 25 percent between 1900 and 1914.[2] This price rise affected the academics in the several countries quite differently. In Britain, Denmark, and Norway salaries in all academic ranks, and of employees too, remained at 1900 levels.[3] In France and Italy, although professorial salaries remained fixed, those of junior faculty, assistants, and employees rose about 20 percent in 1908-10.[4] In Germany, on the other hand, salaries of the professoriate rose at least as fast as prices, and total incomes even faster, while the salaries of assistants increased very little.[5] In Sweden

[1] Carnegie Foundation, *Bull., 2* (1907), 86; Kevles, *Phys. in America*, pp., 161-62, 240-41; Newcomb, *North American Review* (1902), pp. 145-58. An opposite view of the relative circumstances of the American and German *assistant* is given, with supporting data, by Böttger, *Amer. Hochschulwesen* (1906), pp. 49-50.

[2] France, Inst. nat. de la stat., *Ann. stat., 58* (1951), 515.* Cf. the 20 to 25 percent rise in prices for laboratory apparatus and material recorded in Württ., Kammer der Abgeord., *Verhl.*, 39. Landtag (1913), Beilagenband 1, Heft VI, pp. 102-03.

[3] G.B., Board of Ed., *Reports* (1900-01, 1910-11); A. Chapman, "Physics at Oxford" (1973). For Denmark and Norway see note a to Table B.1.

[4] France, Min. ed. nat., personal communication (April, 1973); France, Min. instr. publ., *Recueil des lois, 6* (1898-1909), 1119-21, and 7 (1909-14), 213-15, 636. Pisa, U., *Annuario* (1909-10), reporting the law of 19 July 1909. Salaries of juniors also rose substantially (about 50 percent) in Finland while professorial salaries stagnated; Helsinki, U., *Redogörlse* (1905-08), pp. 129-34, and (1911-14), p. 165.

[5] In the Bavarian universities between 1900 and 1910 the average *salary* increased for ordinary professors from 6,000 to 7,000 marks, for extraordinary professors from 3,460 to 4,460, for assistants from 1,300 to 1,370. At the T.H. Munich the

professorial salaries rose 25 percent, those of junior faculty 50 percent, and the assistants also obtained an increase.[6] The Austrian assistant, relatively well paid, maintained his position, but the upper ranks probably fell behind, without lecture fees certainly well behind, their German counterparts.[7] The Swiss seem generally to have kept pace with the Germans,[8] and also in America the relative position of the academics, at least of those on the high rungs of the ladder, was improving.

Useful as the figures of Table B.1 are in estimating the total investment in physics, they conceal national differences of considerable significance. In Germany, for example, the Ordinarii negotiated their salaries and especially their fringe benefits with the ministries of the several states: a man in demand might emerge from such negotiations with a guaranteed income several times the average.[9] Again, in Germany the amount brought by fees might be very great indeed for a popular professor or for those who, popular or not, controlled the courses necessary for intending engineers, physicians, or secondary school teachers.[10] Students paid 20–30 marks to attend a semester lecture course, and 300 marks for a doctoral degree,[11] so that even in the middle-sized universities, especially in the 1890's, men fortunately placed might easily take 10,000 marks in fees. Naturally one tried to make one's lectures entertaining. When Heinrich Kayser went to Bonn as Ordinarius in 1894 he received a salary of 5,500 marks and almost as much again in lecture and examination fees; he (and/or physics) became

---

corresponding increases were from 5,640 to 7,690, 3,325 to 4,125, and 1,730 to 2,010. Compiled from Bavaria, Landtag, Kammer der Abgeord., *Verhl.*, 33. Landtag, Beilagenbd. 4, Budget 26, pp. 94–98, 106, 166, 180; *ibid.*, 35. Landtag, Beilagenbd. 12, Budget 28, pp. 86-90.

[6] See note a to Table B.1.

[7] Austrian assistants' salaries had risen some 22 percent between 1899 and 1912. Austria, Reichsrat, Haus der Abgeord., *Sten. Protokolle*, 21. Session (1912), Beilage 1287; Hellmer in Brünn, T.H., *Festschr.* (1899), pp. 81–83.

[8] Indications of the Swiss situation are gleaned from Switzerland, Dept. d. Innern, *Schweiz. Schulstat., 1911–12, Text, 4,* 59, 85–86, 153, 187, and from W. Meyer, *Finanzgeschichte der U. Zürich* (1940), pp. 49–52, 59.

[9] *Financial Status* (1907), pp. 74–75.

[10] Jesse, *Educ. Review, 32* (1906), 436; Lehmann, *Phys. Inst. Karlsruhe* (1911), p. 88; Karlsruhe, T.H., *Festschrift* (1950), pp. 69, 123. Küchler, "Phys. Labs. in Germany" (1906), p. 198.

[11] *Minerva: Handbuch* (1911), pp. 10–13. The fee for promotion in the philosophical faculties ranged from 180 marks in Kiel to 355 in Berlin, 300 marks being the mean; but as the student had to deliver 150 to 300 (at Kiel!) copies of his dissertation to the faculty, the cost of the degree was probably about 500 to 600 marks. Also the fees for the *Practica* were substantial. Spielmann, *Handbuch der Anstalten* (1897), p. 83.

TABLE B.1

Salaries and Fees (1000's of Marks)[a]

	Professors		Junior faculty		Privatdozenten		Assistants		Employees
	Range	Avg.	Range	Avg.	Range	Avg.	Range	Avg.	Range
Austria-Hungary									
T.H.	3.9–7.3	(7)		(4)		(1)	1.0–2.3	(1.6)	1–2
Univ.	3.6–7.0	(9)		4.2		(1)	1.0–2.3	(1.6)	1–1
Belgium	5.8–?	(8)	4.1–?	(5)					
British Empire	5.5–25	14	3–8	5		(5)[b]	1.1–2.6	2.0	1–3
France		(12)		(7)				(2)	
Paris	7–17	14	4–10				1.5–4.5	2.5	1.2–5
Depts.	7–13	10	4–6.5				1.5–3.5		
Germany									
T.H.	4–15	8		4.5		(1)	0.6–5	1.8[c]	1–2
Salaries	4–10	6	3–4	3.5				1.5	1–2
Fees		2		1				0.3	0
Univ.	4–37	14	3–10	6	0–3	1.0		1.4	1–2
Salaries	3–12	6.5	1.5–5.5	3.5	0–1.5	.12	0.8–2.5	1.4	1–2
Fees	1–25	7.5	1.5–5	2.5	0–2	.85	0	0	0
Italy	4–10.5	(9)	1.5–4.5	(3)		0.5	1.0–2.6	1.9	1.3–2.5
Japan									
Netherlands	6–13	(8.5)		(5)					
Russia	6.6–?	(10)	4.4–?	(5)					
Scandinavia	4–6.6	6.0	2.2–4.4	3.3			1.2–2.2	1.7	1.1–2.4
Switzerland	3–12	(8)		(4)				(1)	1–2.4
United States	9–25	13[c]	5–14	8[c]	3–8[b]	5.5[c]	0.5–3	2	1–5

[a]Here as elsewhere, figures in parentheses are our guesses. Unparenthesized figures are also our own estimates but have some substantial basis in the sources cited below. Eulenburg, *Akademischer Nachwuchs* (1908), pp. 110–12, 134–35; Stötzner, *Öffentl. Unterrichtswesen* (1901), p. 156; Alsace-Lorraine, *Landeshaushaltsetat* (1900), pp. 108–09.

[b]Privatdozenten: figures for the British Empire and the U.S. in this column refer to Fellows and instructors.

[c]Median, not average.

Sources:

Austria: Hellmer in T.H. Brünn, *Festschrift* (1899), pp. 46, 67–68, gives virtually our only data on professors, and he leaves obscure the consequences of the nationalization of lecture fees in 1896; *ibid*, p. 83, Austria, Reichsrat, Haus d. Abgeord., *Sten. Prot.*, 21. Sess. (1912), Beilage 1287, and *Hochschul-Nachrichten, 11* (1901), 11, 157, for assistants; Eulenburg, *Nachwuchs* (1908), pp. 135, 140–41, for junior faculty.

Belgium: Minerva: Handb. (1911), p. 304.

British Empire: We use figures for salaries and fees for 16 professors and 39 lecturers, assistant lecturers, and demonstrators in physics at English and Welsh universities from G. B., Bd. of Ed., *Reports* (1900–01). We have divided this latter heterogeneous group into "junior faculty" and "assistants" according as the salary is above or below £125 (2,500 marks); when above it is seldom less than £150 (3,000 marks). Scotch, Irish, and Canadian incomes are generally consistent with these: Aberdeen, U., *Calendar* (1902–03), pp. 513, 523; Moody and Beckett, *Queen's Belfast, 2* (1959), 711, 718–19; Edinburgh Report in *Sess. Pap.*, 1903, *53*, 513; Galway, Queen's Coll., *Pres. Report* (1899–1900), p. 25; Glasgow, U., *Calendar* (1900–01), p. 621; St. Andrews, U., *Calendar* (1900–01), pp. 553–56; Toronto, U., *Pres. Report* (1901–02), p. 32. For anecdotal evidence see Lodge, *Past Years* (1931), pp. 153, 157; Eve, *Rutherford* (1939), pp. 39, 47, 50, 56; J. J. Thomson, *Recollections* (1937), p. 124; Grant, *Bragg* (1952), p. 13; Johnstone, "Physics Dept., Dalhousie U." (1971), p. 11.

France: Min. ed. nat., Personal communication (13 Apr. 1973); Min. instr. publ., *Statistique, 3* (1889), 98–99, 100–01, 683, 685, 694, 707. For salaries after 1908 see *idem., Recueil des lois, 6* (1909), 1119–22. For anecdotal evidence see France, Assemblée Nationale, *Débats* (1900), 24 Jan., Berthelot, p. 183, Villejean, p. 188; M. Curie, *Pierre Curie* (1923), p. 32; A. Langevin, *Langevin* (1971), p. 53; E. Curie, *M. Curie* (1937), pp. 238–39; Bouty, quoted in Paris, U., Conseil acad., *Rapports* (1906–07), p. 103. Cf. note 44.

Germany:

Universities: *Financial Status* (1907), pp. 74–82, which includes the fees, and Lexis, *Unterrichtswesen, 1* (1904), 43–44, 47. We include Wohnungsgeld. A complete salary distribution for the three Bavarian universities for 1900–01 is given in the Kammer der Abgeord., *Verhl.*, 33. Landtag, Beilagen-Bd. 4, Budget 26, pp. 94–98, 166; for U. Giessen for 1897 in Hesse, Landstände, 2. Kammer, *Verhl.*, 30. Landtag, Beilagen-Bd. 1, Beilage 92, Anlage, p. 19. Further information is in *RIE, 22* (1891), 231–32; Technische Hochschulen: For income including fees, *Financial Status* (1907), pp. 82–84 (which averages, based upon data from the four most important schools—Aachen, Berlin, Danzig, Hannover—are somewhat too high); for fees, Prussia, Stat. Landesamt, *Stat. Jahrbuch, 1* (1903), 170; Damm, *T. H. in Preussen* (1899), pp. 57n, 75n. Salary distributions for T.H. München (1900) and T.H. Darmstadt (1897) are given in the parliamentary papers cited above for the Bavarian and Hessian Universities, pp. 106, 180, and 19, respectively. For anecdotal evidence see Lexis, *op. cit., 4*, 296; Lehmann, *Phys. Inst. Karlsruhe* (1911), pp. 84–87.

Italy: Haguenin, *RIE, 35* (1898), 335n.; Italy, Min. pubbl. istr., *Bolletino, 7* (1881). The average remuneration of the Italian Privatdozent is constructed from the lists of the same in *Minerva: Jahrb.* together with *quote* paid liberi docenti in the faculty of science and mathematics at U. Turin, 1895–97: *Annali* (1898), p. 87. For a detailed, and very lengthy, list of posts and salaries acquired in one career see: Polvani, *Pacinotti, 2* (1934), 1034–35.

Netherlands: Minerva: Handb. (1911), pp. 161–63.

Russia: Minerva: Handb. (1911), p. 369; Darlington, *Education in Russia* (1909), pp. 260–63.

Scandinavia: Copenhagen, *Årbog* (1907–08), pp. 35–36. Helsinki, U., *Redogörlse* (1905–08), pp. 129–34, 176; (1911–14), p. 165. Oslo, U., *1811–1911, 1* (1911), 286–87, 360; *1911–1961, 2* (1961), 236; personal communication (21 Mar. 1973), reporting from Oslo, U., *Årsberetning*, and Norway, Storting, *Proposijon.* Lund, U., *Historia, 4* (1968), 187. Uppsala, U., *Festskrift, 1* (1897), 276. Henriques, *Skildringar, 2* (1927), 272, 353, 423. Sweden, Utbildningsdepartement, personal communication (27 Mar. 1973), reporting from *Sveriges Statskalendar* and Statsliggaren. Also *Minerva: Handb.* (1911), pp. 172, 176–77.

Switzerland: Meyer, *Finanzgesch. U. Zürich* (1940), pp. 49–52, 59; Basel (Canton), *Staatsrechnung* (1900), p. 13; Neuchâtel (Canton), *Décret portant budget* (1900), pp. 50–51; (1910), p. 58; Geneva (Canton) *Budget* (1900), pp. 27–31; (1910), pp. 29–31; Dept. d. Innern, *Schweiz. Schulstat., 1911–12, Text, 4* (1915), pp. 59, 85–86, 153, 187; *Minerva: Handb.* (1911), p. 148.

United States: Financial Status (1907), pp. 20–31, gives ranges and averages for professors in all fields in the 102 wealthiest institutions in 1907; our averages are the median average salaries in the top 30 institutions. For anecdotal evidence see Erikson, "U. of Minnesota Dept. of Physics" (n.d.), pp. 14–17; Eve, *Rutherford* (1939), pp. 81, 125–27; Reingold, *Sci. 19th C. America* (1964), pp. 314, 318–19; Birge, "Physics Dept. Berkeley," *1* (1966), III (15–16), V (8); Solberg, *U. Illinois* (1968), pp. 257–58; Gray, *U. Minnesota* (1951), p. 106; Böttger, *Amer. Hochschulwesen* (1906), p. 49.

popular, and in a few years his income from fees had risen to more than 20,000 marks.[12] Clearly, great disparity in income might exist between the experimentalist who gave the large introductory courses and the theoretician condemned to advanced subjects; between the famous and the routine practitioner; and between the university man and his colleague in the Technische Hochschule, where usually only a small portion of fees went to the instructor.[13]

In the Anglo-Saxon countries the range of *regular* income—salary plus fees, if any—for full professors was smaller than in Germany. In Britain, for example, the salary of the best paid professor was roughly twice the average and four times the lowest, the 5,300 marks paid a beginning professor in the impoverished university colleges of Wales. British university students paid very high fees—often more than five times the German[14]—

[12] Kayser, "Erinnerungen" (1936), p. 180. At provincial Giessen in 1888 Röntgen was already drawing as much in fees, 5,500 marks, as in salary. Glasser, *Röntgen* (1934), pp. 97-98. Six years later, at Würzburg, he had 6,360 in salary and 10,540 in fees, largely from medical students. (Bayerisches Hauptstaatsarchiv, Munich, Mk 17921* "Röntgen," Blatt 14.)

[13] The maximum fees allowed a Prussian T.H. professor were 3,000 marks. Lexis, *Unterrichtswesen, 4,* Pt. 1 (1904), 33, 212; Damm, *T.H. Preussens* (1909), pp. 262, 268-72; Scheffler, *T.H. und Bergakademien* (1893), pp. 5-7. Reforms took away part of the fees of university Ordinarii appointed after 1897, viz. one-half of fees in excess of 3,000 marks (4,500 in Berlin); the money so extracted went into a fund to help Privatdozenten and others, but not to the institutes. Lexis, *Unterrichtswesen, 1* (1904), 45; Breslau, U., *Chronik* (1908-09), p. 43. In 1896 Austria nationalized lecture fees and set salaries by seniority. Busch, *Gesch. Privatdoz.* (1959), p. 97; Vienna, U., *Inaug. des Rektors* (1899), pp. 40-43. In the Swiss universities the instructors continued to receive fees without limitation German style and to level student fees of c. 4 marks per semester hour (c. 1 mark at the ETH, Zurich). Switzerland, Dept. d. Innern, *Schweiz. Schulstat., Text,* Pt. 4 (1915), passim; *Minerva: Handbuch* (1911), pp. 148-57. Russian university students paid fees to their instructors at about two thirds the German rate, i.e., c. 100 marks per student per year on the average. Darlington, *Educ. in Russia* (1909), p. 263. In the Netherlands the students at the three state universities paid a fixed annual tuition of 350 marks (700-1,500 marks for the doctorate), in which the instructors apparently did not share, while at Amsterdam the students paid substantial lecture fees. Utrecht, Universiteit, *Jaarboek* (1905-06), pp. 145-46; *Minerva: Handbuch* (1911), pp. 161-63. Similarly in Italy and Scandinavia the instructors apparently received no fees, which were low in the north (*ibid.,* pp. 172, 176-77) and high in the south (*ibid.,* p. 321): 50 marks for matriculation, 100 for inscription, 15 for examination, and 80 for diploma in Italy in 1900, which became 50, 390, 65, and 80, respectively, in 1902. Turin, U., *Università di Torino* (1900), pp. 222-23; Rome, U., *Annuario* (1902-03), p. 257.

[14] British lecture fees were about 60 marks per year per class hour per week against 10 marks in Germany; British laboratory fees were about 300 marks per year for one day per week. Birmingham U., *Calendar* (1900-01), p. 141; G.B., Board of Ed., *Re-*

but the rates were not uniform and the proportion as well as the amount of fees retained by the British professor varied from one institution to another. In Rayleigh's day the Cavendish fees all belonged to the director; he plowed his back into the laboratory, a practice followed by his successor, J.J. Thomson, who built a new wing with accumulated student fees.[15] Elsewhere, Oliver Lodge received two-thirds of his fees at Liverpool, and Poynting retained one-quarter of his at Birmingham.[16] Poynting's situation was perhaps typical of the British professor of 1900, who appears to have drawn between 3,000 and 6,000 marks annually from fees.[17] In the United States fees customarily went into the general fund of the university. The professor's regular income was his salary alone and its average value did not vary by more than a factor of two among the institutions in our set.[18]

In France the professor did not receive fees. He improved his regular income via the *cumul,* which brought some professors of physics above 20,000 marks a year. Among those who did well at this typically French practice we may mention G. Lippmann, professor of physics and laboratory director at the Sorbonne and secretary of the Bureau des Longitudes, and E. Mascart, the government's favorite physicist, professor and secretary of the faculty at the Collège de France, director of the Meteorological Bureau, president of the council of the École supérieure d'Électricité, and member of the Bureau of Weights and Measures.[19]

In most countries the Ordinarius received salary supplements and fringe benefits which we have not included in Table B.1. Among supplements the most important was an allowance given either as a contribution toward house rent (and included in Table B.1) or in kind, as a home provided at

---

ports (1900-01), pp. 281-82, for U. Manchester; A. Chapman, "Phys. at Oxford" (1973), citing Oxford, U., *Gazette*; Liverpool, U., *Calendar* (1902-03), pp. 166-67. At University College London the annual "composition" was an elegant 101 guineas. *Calendar* (1909-10), p. 299.

[15] Rayleigh, *Rayleigh* (1968), pp. 102-03, and *J.J. Thomson* (1942), p. 46.

[16] Lodge, *Past Years* (1931), pp. 153, 157; G.B., Board of Ed., *Reports* (1900-01), pp. 5, 27; *ibid.* (1894-95), p. 88.

[17] Moody and Becket, *Queen's Belfast* (1959), *2*, 711, 718-19; Galway, U., *Pres. Report* (1899-1900), p. 25.

[18] The fixed maximum of American salaries, unnegotiable like the German and unimprovable by fees, were a constant cause of complaint; see, e.g., *Financial Status* (1907), pp. 23, 74. If we had included the smaller American colleges, the salary schedule would have been much bleaker, descending to below 2,000 marks per year. Instructors at Harvard were paid better than full professors at some schools. Harvard U., *President's Report* (1905-06), p. 17.

[19] *Minerva: Jahrb.* (1900). For Mascart see Langevin, *Revue du mois, 8* (1909), 385-406.

low or nominal rent. The practice was best established in Germany, where the low (or no) rent residence might be built right into the institute.[20]

The most important fringe benefit was the right to a pension. Here the Anglo-Saxon physicists were far behind their continental counterparts. The German Ordinarius held his chair and most of his salary for life; he had nothing to fear but the loss of fees if he ceased lecturing, and his widow and children received pensions after his death.[21] The French or Italian professor retired with an adequate stipend and his kin were entitled to state support.[22] But there was no general pension plan for university faculty in Britain or her colonies, and the Royal Society's tiny relief fund for needy scientists made a mockery of the plight of the academic scientist.[23] British professors perforce devoted much of their energy to providing for their old age. The situation in the United States was no better before the establishment of the Carnegie pension scheme.[24]

To return to Table B.1: as one descends through the academic ranks, one finds that the percentage range in regular income (as well, of course, as its absolute value) decreases at every step. The most homogeneous group, the regular (as opposed to occasional or part-time) assistants at the bottom of the ladder, received much the same starting salary in every country, 1,000 to around 2,000 marks. What that might buy will occupy us momentarily. The finances of the Privatdozent are not so easily categorized. By right of his position he was entitled only to his students' fees which, since he did not give introductory courses and did well to avoid drawing students from the lectures of the chairholder, brought an average of 850 marks per year, and almost never more than 2,000 marks.[25] A few Privatdozenten, at most 10 percent, managed to obtain small stipends,

[20] See below, section D.

[21] *Financial Status* (1907), pp. 87ff.; Lexis, *Unterrichtswesen, 1* (1904), 48–49.

[22] France, Min. instr. publ., *Recueil des lois,* 7 (1909–14), 649–50.

[23] MacLeod, *Technology and Society, 6* (1970), 47–57. Cambridge set aside only 4,000 marks per year for the pension fund for 44 professors. *Nature, 75* (1907), 404. Cf. Bragg (in Australia) to Rutherford, 2 August 1907: "You mention a probable minimum of 3,500 dollars for a chair of mathematical physics at McGill, with the right to a Carnegie pension. Here I receive £800 a year, but have no claim to a pension." (Rutherford Papers, Cambridge Univ. Libr.)

[24] Only 5-6000 teachers in 73 institutions (none state-supported) in the United States and Canada belonged to the scheme by 1915, when a new system was introduced. Carnegie Found. Adv. Teaching, *Bulletin*, No. 1 (1907), and Pritchett, *Plan of Insurance* (1916), pp. 53–58. For pensions at the best schools, Thwing, *North American Review, 181* (1905), 725–26.

[25] *Financial Status* (1907), p. 68; Eulenburg, *Nachwuchs* (1908), pp. 110–15; Busch, *Gesch. Privatdoz.* (1959), pp. 117–26.

amounting, in the Prussian system, to 1,200 marks.[26] But one could scarcely count on such largesse, and many Privatdozenten supplemented their income by taking assistantships. In 1907 about 45 percent of the Privatdozenten and about 30 percent of the Extraordinarii in the natural sciences were teaching or laboratory assistants.[27] The situation seems to have been much the same in other countries with German academic ranks, like Italy, where, however, the government paid the libero docente a salary proportional to the number of students he taught.[28]

We have included in the Privatdozent column an estimate of the amount available to British physicists through prize fellowships at Oxford and Cambridge. Reforms introduced in 1882 transformed these fellowships (not to be confused with those held by professors or college officials in virtue of their positions) into research posts, won by open competition and tenable for six or seven years, at a stipend of some 6,000 marks p.a.[29] The number held by physicists varied: we have used the figures for 1914, the only secure ones we have (Table A.4), in computing the total fellowship support for 1900.

b. Other Sources of Income

We can do little more than list the most important sources of irregular and extra-professional income: parental allowances and personal wealth, scholarships, moonlighting, fellowships and prizes, honoraria for publications and lectures, royalties on books and patents, fees for expert advice and services. Although in the following we consider a few of these sources, only in the case of prizes and British prize fellowships can we estimate the amounts in question. The yields from every other source remain unknown and have been ignored in Table I.

Estimates of parental allowances or personal wealth would require separate and extended investigation of the social origins of the academic

[26]For up to five years. Eulenburg, *Nachwuchs* (1908), p. 112. Prussia gave only 60,000 marks, which had to be divided among 528 docents in 1900. Prussia, *Statistik, 193* (1905), 6. Evidently less than one in ten received a substantial stipend; cf. Busch, *Gesch. Privatdoz.* (1959), pp. 110–13.

[27]Eulenburg, *Nachwuchs* (1908), pp. 58, 65; Ferber, *Lehrkörper 1865–1954* (1956), pp. 87–88, 102; Lexis, *Unterrichtswesen, 1* (1904), 46–47; Busch, *Gesch. Privatdoz.* (1959), pp. 108 ff.

[28]Haguenin in U.S., Office of Ed., *Reports* (1897–98), 1453; *RIE, 31* (1896), 599–600. We assume that the Austrian universities also paid their Privatdozenten in proportion to numbers taught.

[29]Howard, *Finances of St. John[s]* (1935), pp. 218–19; Ward, *Victorian Oxford,* (1965), p. 310; Winstanley, *Later Victorian Cambridge* (1947), pp. 347, 354–55.

physicists in the several countries. Although of fundamental importance, in particular for our calculations of research productivity in Section E, it is beyond the scope of this paper. We would guess that parental allowances and personal wealth contributed most in Germany and in the United States —in the former because the prestige of the academic career was so high, in the latter because it was so low.

In the matter of moonlighting and royalties we can say little more. Frenchmen, as we have seen, often added nonacademic jobs to their *cumuls*. Physicists in all nations received fees for additional examining work, for special government service, for appearing as expert witnesses, and for consultation—or even collaboration—with industry. Anglo-Saxon physicists probably led the world in extracting money from this last source. Despite the close relations between German science and technology, the German professor of physics did not rush to make money from his research: "It is well known throughout the world that the physical laboratories of Germany have no windows looking towards the patent office."[30] German physicists who became involved in business often left their university positions for industrial jobs.[31]

Not so the British physicist: Lord Kelvin, the holder of many lucrative patents, ran an instrument-making firm which, in many respects, was a branch of his laboratory, and which he visited for an hour or so every day.[32] Other English academic physicists and many American ones often acted as consultants and expert witnesses for electrotechnical firms, electric streetcar companies, etc.[33] An interesting illustration of national differences in this matter is afforded by the radium industry. Marie Curie prided herself on eschewing the commercial exploitation of radium; F. Giesel manufactured it at the Braunschweig Quinine Works and sold it

[30] Münsterberg, *Science, 3* (1896), 162, à propos of Röntgen's discovery. Also quoted in Glasser, *Röntgen* (1959), p. 277. Cf. Glasser, *Dr. W.C. Röntgen* (1958), p. 125; Nitske, *Röntgen* (1971), pp. 194–95; and Poincaré, *Foundations of Sci.* (1929), pp. 279, 355.

[31] E.g., Abbe and Straubel (Zeiss) and Raps (Siemens and Halske). Among exceptions to this rule may be mentioned W. Gaede, who made a fortune from his vacuum pumps without leaving his Hochschule. H. Gaede, *Wolfgang Gaede* (1954), p. 55.

[32] Gray, *Nature, 55* (1897), 490; Thompson, *Kelvin, 2* (1910), 717–18, 994, 1155. Compare the indifference of Pierre Curie to the commercial exploitation of his electrometer. M. Curie, *Pierre Curie* (1928), pp. 48, 63–64, 95–96, 110–11.

[33] Eve, *Rutherford* (1939), p. 65; Larmor, ed., *Stokes, 1* (1907), 252–53, 257–58; Lodge, *Past Years* (1931), pp. 147–51, 175 ff.; Williams, *Morley* (1957), p. 211; J.S. and H.G. Thompson, *S.P. Thompson* (1920), pp. 116–18, 253, 258–59; Millikan, *Autobiog.* (1951), pp. 133–41.

at cost; W. Ramsay set up a corporation which hoped to refine and market it for profit.[34] We take it that the relative crassness of British and American physicists was related to their need to supply for themselves what the European received through his pension system.

Royalties received from publication amounted to substantial sums for the few authors of successful textbooks; on the average they contributed little to gross income.[35] Nonetheless publishing brought welcome small change, and we can only commend a system which paid not only for popular, but even for scholarly articles. German scientific journals paid between 20 and 40 marks per sheet (sixteen printed pages octavo).[36] *Nature* paid about 7 marks a column, slightly more than the rate at which *Naturwissenschaften* proposed to reward its contributors.[37]

Finally we may mention prizes given by learned societies as rewards for past achievement. From Table C.4 it appears that the French were the most generous in what we might call scientific philanthropy. In fact French predominance here is the obverse of a serious shortcoming: the prize money amounted to roughly nine times the grants for projected researches, a proportion just about reversed in other countries.[38] Some of this prize money did find its way into research, but much of it probably remained in the pockets of its recipients. Hence we have treated it as income. In Section E we guess how much of it became available for research.

2. Academic Careers

In Table B.2 we give the average age at which physicists reached the various ranks, and began to earn the corresponding salaries given in Table B.1. Graduation signifies taking the A.B. or B.S. in the United States and Britain, the Agrégation in France, the Ph.D. in Germany; "graduate

[34] Tilden, *Ramsay* (1918), p. 169; Hahn, *Autobiography* (1966), pp. 38-39; Howorth, *Soddy* (1958), p. 99; Rutherford, *Radioactive Substances* (1913), p. 17; *Le Radium,* (1904), passim. M. Curie did have the use of a small factory for processing minerals courtesy of A. de Lisle, an industrialist who supported her research in other ways as well. Paris, U., Conseil, *Rapport* (1912-13), p. 195; E. Curie, *M. Curie* (1937), pp. 199-200.

[35] *Financial Status* (1907), p. ix.

[36] Kirchner, *Das Deutsche Zeitschriftenwesen, 2* (1962), 472; M. Planck to Fr. Vieweg & Sohn, 3 December 1892 (Vieweg Archiv, Braunschweig), re 40 marks/sheet paid by the *Jahresbericht über die Fortschritte der Physik.*

[37] MacLeod, *Nature, 224* (1969), 443; Berliner and Thesing, circular letter of October 1912 (copy in archive of Springer Verlag, Berlin).

[38] Levy, *CR, 131* (1900), 1040.

degrees" means Docteur-ès-sciences in France, Ph.D. in the United States, and Habilitation in Germany. (We recognize that the American A.B. and German Ph.D. are scarcely comparable degrees and that the German Habilitation is not a degree at all, but a process qualifying holders of the Ph.D. to teach in universities and Technische Hochschulen. We have

TABLE B.2

Median Age of Entry into Physics Posts[a] (with range and size N of sample)

	Graduation			Post-grad. degree			Jr. faculty			Full prof.		
	Age	Range	N	Age	Range	N	Age	Range	N	Age	Range	N
British Empire	22½	20–26	44				26½	22–38	57	32[b]	22–50	81
France[c]	23½	21–28	34	28½	33–45	63	30	22–47	49	36½	25–53	60
Germany												
Univ.	24	20–32	74	27	20–50	91	32½	25–43	61	37½	25–67	77
T.H.							33	28–46	12	34½	25–56	30
Italy							34	32–34	4	32	25–35	28
United States[d]												
Leading schools	22	18–28	72	27½	23–43	48	26½	20–37	53	33	23–48	52
Other	22	18–29	39	28	22–43	35	25½	21–33	26	30	20–44	40

[a]For France, Germany, and the U.K. this table is based on samples covering more than three-quarters of the academic physicists active in each country in 1900, plus a few of the better-known men who entered the field between 1900 and 1910; no data on positions entered after 1914 are used. Names come from *Minerva: Jahrb.*, from course lists in *Phys. Zs.*, from *Commonwealth Universities Yearbook*, and from university calendars. For the U.S. we use the men *Minerva: Jahrb.* lists at our set of leading schools, supplemented by the prominent (starred) physicists in Cattell, *Amer. Men of Sci.* (1910), p. 582. Both sources tend to miss men in the lowest ranks in 1900; accordingly our American sample refers to a group of men a few years older, on the average, than the British, French, and German samples. The Italian sample is men holding posts in science faculties in 1903, from Italy, Min. pubbl. istr., *Annuario generale univ.* (1903).

[b]The median age of reaching a first professorship outside the United Kingdom was 32½ (23 to 55, N = 38).

[c]The median age of reaching the first professorship in the major Parisian institutions alone (Sorbonne, Collège de France, École normale, École polytechnique, Muséum d'histoire naturelle, and the Bureaus of Weights and Measures and Meteorology) was 40½ (26 to 54, N = 27), while for all other schools it was 35 (25 to 50, N = 38).

[d]"Leading schools" enters the data for all men in the schools in our set, whether or not starred by Cattell; "other" signifies starred physicists in other schools. All biographical data are from Cattell, *Amer. Men of Sci.* (1906). Junior faculty includes instructors, who evidently often took their positions before they received the Ph.D.

organized our data in this way only because it facilitates multinational comparison of holders of academic positions.)

One sees that the typical man had graduated in his early twenties, finished his research degree five or six years later, and entered into his first decently paid position at about the age of thirty. He won a full professorship before he was forty, and there stopped, unless he changed schools or took a post in the university administration. Note that in Britain, where there existed no advanced degree quite equivalent to the time-consuming research degree found elsewhere, the typical physicist of 1900 had reached both a secure junior post and a professorship some six years earlier than his continental counterpart. Our sample of senior American physicists shows a similar precocity.

Inspecting the rungs of the job ladder, on the lowest we find the research student, who devoted himself to his own research project and, in the United States, to advanced course work.[39] Often the research student paid a laboratory fee, if only the 15 marks per semester at Leipzig, to earn the right to contribute his mite to science.[40]

A proportion of research students—about one in three at the Cavendish, two in five at Leipzig in 1900[41]—were also assistants. In France, Germany, and the United States about half of all men who eventually held faculty posts in physics had been assistants before taking a postgraduate degree.[42] For his official duties, which usually occupied no more than 50 percent of his time,[43] the assistant received a salary close to the prevailing wage for

[39]Cf. Trowbridge in Harvard U., *President's Report* (1894-95), pp. 207-08; Webster in *Clark U. 1889-1899*, pp. 90-92.

[40]Küchler, "Phys. Labs. in Germany" (1906), p. 192; Wiener in Leipzig, U., *Festschrift, 4*, Pt. 2 (1909), 69. In the Paris faculty of science in 1901, 51 of 94 advanced students in all fields paid fees. Paris, U., Conseil, *Rapport* (1901-02), p. xxvi.

[41]The ratio of independent researchers to assistants at Leipzig was 2/2 in 1873, 4/3 in 1887, 6/4 in 1900, 18/10 in 1910. Wiener in Leipzig, U., *Festschrift, 4*, Pt. 2 (1909), 57.

[42]Counts from the samples defined in Table B.2. For France 38/86 = 44 percent, for Germany 54/100 = 54 percent, and for the United States 31/89 = 35 percent are known to have held assistantships; the true proportion must have been still higher.

[43]At the Cavendish one might "demonstrate" three days a week. Cavendish Lab., *History* (1910), p. 90; Eve, *Rutherford* (1939), p. 39. For the situation at Manchester, Heilbron, *Moseley*, p. 47. Circa 1905 Joffe's duties in Röntgen's lab occupied him only two hours each afternoon. Joffe, *Begegnungen* (1967), p. 32. In 1915 Wegener was obliged to be in the Marburg Institute from 9 to 1 and from 3 to 6. E. Wegener, *Alfred Wegener* (1960), p. 148. See Glum, *Wiss., Wirt. und Politik* (1964), p. 342. Assistants who were Privatdozenten would also have to lecture three to eight hours a week. Tompert, *Lebensformen* (1969), p. 28. Cf. Küchler, "Phys. Labs. in Germany" (1906), pp. 189-90; Weinberg, *Phys. dans 206 lab.*; and *Financial Status* (1907), p. 31.

unskilled labor; it was scarcely sufficient to support himself in middle-class style, and certainly inadequate to support a wife.[44] The fact that assistantships often were held by undergraduates, especially in the smaller, provincial schools, points to a shortage of qualified assistants, and suggests that the lack stemmed largely from want of financial incentives.[45] The financial plight of the scholarship student was about the same as that of the assistant.[46]

The freedom of research these young men enjoyed varied greatly from country to country and from one establishment to another. In Britain the professor had the power to control the subjects of research in his laboratory; whether he chose to exercise it, and to exercise it to concert an attack on a particular class of problems, depended primarily upon his personality, policy, and research strategy. In Paris doctoral researches quite often bore no connection with the research interest of the man in whose laboratory they were carried out.[47] The French enjoyed comparing this *liberté* with the situation in Germany[48] where, to quote Max Weber, the assistant was a "quasi-proletarian" in a "state-capitalist enterprise."[49]

[44]Böttger, *Amer. Hochschulwesen* (1906), pp. 49-51, on the plight of the German assistant; Austria, Reichsrat, Haus der Abgeord., *Sten. Protokolle,* 21. Session (1912), Beilage 1287, on that of the Austrian; Bouty in Paris, Conseil de l'U., Conseil acad., *Rapports* (1906-07), pp. 101-05, on that of the assistants at the Sorbonne. The upper end of the range of salaries attributed to French assistants in Table B.1 applies only to the chefs des travaux practiques. These held permanent, full-time positions with rather obscure, partially administrative, duties. The chefs des travaux were seldom active in research, and the majority never earned the doctorate, but they nonetheless often advanced to higher administrative or faculty positions.

[45]J.J. Thomson to R. Threlfall, 14 Dec. 1893 (Thomson Papers, Camb. Univ. Library): "There is such a demand now for physicists that I shall be at my wit's end to get demonstrators for next term, as my old ones keep going off to new posts." Cf. note A.22 and Gerbod, *La condition universitaire* (1965), pp. 588-89, for the relative attractiveness of secondary school posts over assistantships. Likewise Hellmer in Brünn, T. H., *Festschr.* (1899), p. 81.

[46]The "1851 Exhibitions," paying 3,100 marks per year, were perhaps the fattest; in France, as late as 1912, the doctoral scholarships at the Sorbonne (some 11 for all the sciences) paid only c. 1,200 marks each. Great Britain, Board of Ed., *Reports* (1900-01), pp. 7-8, 117; Eve, *Rutherford,* p. 47; France, Min. instr. publ., *Enquêtes, 108* (1914), 91-93.

[47]For example, a number of dissertations on X and cathode rays, in which Lippmann never displayed much interest, were prepared in his laboratory. His attitude was distinctly libertarian. Paris, Conseil de l'U., Conseil acad., *Rapports* (1902-03), p. 109.

[48]Paul, *Sorcerer's Apprentice* (1972), pp. 16-17.

[49]Weber, *Essays* (1946), p. 131. The view of the university institute in the natural sciences as "ein capitalistisches Seitenstück zu den fabrikativen Grossbetrieben" was a commonplace in this period. A. Prokop, *Rektoratsrede* at T. H. Vienna, 1896, p. 13, quoting A. Wagner, *Rektoratsrede* at U. Berlin. Cf. Busch, *Gesch. Privatdoz.* (1959), pp. 69-72.

Foreign physicists sometimes admired the disciplined research army of the German professor.[50] Its value depended upon its commander. A Nernst might order his to extremely fruitful projects (like testing Einstein's theory of specific heats),[51] while a Karsten might divert an entire institute into routine meteorological work.[52]

The first step up the academic ladder, into a junior faculty position, depended very much upon the support of the local chairholder. The danger of this situation may best be illustrated in Germany, where about half the Privatdozenten in the philosophical faculties of the large Prussian universities habilitated and taught in the same university in which they had earned their doctorate;[53] they consequently often knew only a single Ordinarius, who had the power not only of directing their research, but also of controlling the publication of their results.[54] The system was wasteful, unfair, and frustrating. After years of financial hardship the unlucky Privatdozent might find himself no closer to a chair, and indeed 44 percent of a sample of 136 physicists who habilitated between 1890 and 1909 never got one at all.[55] Moreover, as one sees from Table B.3, in the leading countries those who did receive a call had to wait longer for it as time went on, for the number of junior men (including Privatdozenten) grew much more quickly than that of the chairs.[56]

Assuming he had awakened the sympathies of his Ordinarius, the

[50] Merz, *European Thought, 1* (1964), 67. Cf. Cardwell, *Sci. in England* (1957), pp. 106, 126, 170-71.

[51] G.P. Thomson, R. Soc. Lond., *Biog. Memoirs, 4* (1958), 45; Küchler, "Phys. Labs. in Germany" (1906), p. 193.

[52] Schmidt-Schönbeck, *Phys. Kieler U.* (1965), pp. 81, 125; cf. Hertz, *Erinnerungen* (1927), passim, for his difficulties at Kiel. The opposite extreme of "great, often too great freedom" characterized Warburg's important institute in Berlin. Pohl, *Phys. Bl., 28* (1972), 543.

[53] Lossen, *Anteil der Katholiken* (1901), pp. 126, 130, referring to 1884 to 1897. In the same period 40 percent of the philosophical faculty in the Prussian universities reached the Ordinariat via advancement within their own university (*ibid.*, p. 138). For inbreeding in American Physics faculties, which reached (and indeed surpassed) the Prussian level only at MIT, see Cooke, *Acad. Efficiency* (1910), pp. 77-78.

[54] See, e.g., Röntgen, *Phys. Zs., 5* (1904), 168; Glasser, *Röntgen* (1959), pp. 96-97; and Röntgen, *Briefe an Zehnder* (1935), pp. 6, 8-10. We suppose that, in principle, prior approval of publication had also to be obtained from institute directors in Britain. Cf. Moseley to his sister, 7 April 1912, in Heilbron, *Moseley* (1973), p. 187: "I have just finished my β ray paper, and wait till Rutherford comes back from his holiday to publish it."

[55] Ferber, *Lehrkörper, 1865-1954* (1956), p. 85; the corresponding fraction for those habilitating between 1850 and 1869 is 33 percent. About half of the Privatdozenten in the philosophy faculties in 1884 were still Privatdozenten in 1896. Cf. *Financial Status* (1907), p. 68.

[56] Supra, Table A.2. Cf. Harvard U., *President's Report* (1905-06), pp. 14-21.

TABLE B.3

Average Age at Time of Reaching First Professorship

	First Call 1880–1889	First Call 1900–1909
British Physicists[a]	30	37
U.K.	30	38
Other	27	34
German Scientists[b]	37	40
Italian Scientists[c]	32	38
Scandinavians[d]	38	39

[a]From the sample of physicists defined in Table B.2, but including some men who died before 1900.

[b]Ferber, *Lehrkörper, 1865–1954,* p. 132.

[c]Italy, Min. pubbl. istr., *Annuario gen. univ.* (1903). The sample is all ordinary professors in 1903; the second figure is for men called to their first chairs in 1898–1903.

[d]Average age of the professorate in all faculties at the universities of Copenhagen (38.5, 39.3) and Oslo (36.6, 38.8) in the years 1879, 1906 and 1877, 1901 respectively. Copenhagen, *Aarbog* (1907–08), p. 12; Oslo, U., *1811–1911, 1* (1911), 359.

Privatdozent after a few years of lecturing would begin his advance by receiving the *title* of extraordinary professor. Often the title brought no salary with it. His first permanent salaried position was usually an Extraordinariat, which, under the prevailing "monarchical constitution" of the institutes and aristocratic composition of the faculties,[57] brought financial but not professional independence. He might, for example, be appointed to lecture on the topics the Ordinarius preferred not to treat, and the terms of his appointment would generally explicitly subordinate him to "the Baron of the Institute."[58] And, of course, he had no guarantee that he would ever advance further.

[57]Lamprecht, *Zwei Reden* (1910), p. 40; Waentig, *Zur Reform* (1911), pp. 2, 10–14. For the "Governance of the University" and the "Dominion of the Ordinarius," see Forman, *Phys. in Weimar Germany* (1968), pp. 80–100.

[58]The phrase is Willstätter's, *From My Life* (1965), p. 297, where translated as "lord." See Lummer, *Festschr. U. Breslau, 2* (1911), 444, and M. Planck to Prussian Kultusminister, 21 October 1913, in Planck, *Schriftstücke,* Nr. 13. E. Meyer's letter of appointment as ausserordentlicher Professor, Tübingen, 8 February 1912, stated that Meyer "was informed when offered the appointment that he was supposed to cooperate with the professor of experimental physics, supporting him and, when necessary, substituting for him." (U. Archiv., Tübingen.) The Nichtordinarien organized in 1908 to try to improve their lot: Busch, *Gesch. Privatdoz.* (1959), p. 112; Paulsen-Lehmann, *Gesch. gelehrt. Unter., 2* (1921), 708–10; Ben-David, *The Scientist's Role* (1971), p. 130.

In France the junior man was likewise dangerously dependent upon his superior for both initial placement and advancement. Furthermore, junior faculty often held positions, like the suppléance, under poorly defined and uncertain conditions.[59] The British junior man felt himself in a better position, or at least we do not find the same complaints about subordination to the professor as we do on the continent. His feeling must in part be referred to the College Fellowship, which conferred a measure of independence on roughly half of the better British physicists early in their careers.[60] Their relative security may help to account for the fact that scarcely any Britishers interrupted their careers to teach in secondary schools, as did at least eight of ninety-five Germans and thirty-six of ninety-four Frenchmen in our samples of men who eventually taught physics at universities.[61] A large fraction of American physicists also taught school at some point early in their careers.[62]

We rise now to the professorship. The British method of appointment was in principle, and often in practice, the most open. Vacancies were advertised, and any interested party might apply, armed with whatever testimonials he thought might carry weight. A committee, usually composed of electors specified in the regulations governing the chair, picked the winner on the strength of his testimonials, research achievements, and teaching ability.[63] Patronage played a bigger part in France. There the central educational bureaucracy made appointments on the basis of reports, particularly those of the chairholder under whom the aspiring physicist worked. A few influential Paris professors controlled appointments so closely that provincial professors often had no say in choosing their own assistants.[64] In Italy professors were selected after open competi-

[59] T.N. and P.P. Clark, *Rev. franç. sociol., 12* (1971); de Forcrand, *RGS, 8* (1897), 613; Burdeau, quoted in *RIE, 23* (1892), 170-71; France, Min. instr. publ., *Recueil des lois, 5* (1898), 404.

[60] 14 of the 29 on whom we have substantial information (culled from Royal Society London, *Biographical Memoirs* and *Obituary Notices,* and from biographies).

[61] See Table B.2 for definition of samples, and Tables A.7 and A.8 for the relatively larger proportion of active physicists in secondary schools on the continent. The 36 Frenchmen include 23 normaliens, exactly half the normaliens in our sample; in addition, 9 of our 13 polytechniciens are known to have served in one of the engineering corps. These facts may help explain the Britisher's relatively early entry into a professorship.

[62] Joncich, *American Quarterly, 18* (1966), 667-85, finds the proposition true of about 1/3 of American physicists born in the latter part of the nineteenth century.

[63] Eve, *Rutherford* (1939), pp. 19, 51-52, 78-79; Thompson, *Kelvin, 1* (1910), 161-88; Rayleigh, *Rayleigh* (1968), pp. 99-102; Howorth, *Soddy* (1958), pp. 51-53, 191-93.

[64] Caullery, *Revue du mois, 4* (1907), 532; T.N. and P.P. Clark, *Rev. franç. sociol., 12,* (1971), Paul, *French Historical Studies* (1972), 7, 19-39, 423-39; Miller, *Physics Today,* Dec. 1966, pp. 47-53.

tion (*concorso*) by a panel of peers appointed by the government; these *concorsi*, intended to offset local influence, were criticized for encouraging the narrow specialization from which Italian physics was said to suffer.[65] Scandinavia likewise had open competition: the faculty chose from a short list compiled by a peer panel, and the government usually ratified the choice.[66] The German call was an elaborate ritual culminating in those delicate negotiations with interested ministries to which we earlier referred.[67]

The professor had reason to be proud of himself. He had outdistanced most of his fellow graduate students. His 10,000 marks a year placed him in the upper bourgeoisie: in Prussia, for example, less than 1 percent of the population had incomes in excess of 9,500 marks in 1900.[68] From this happy conclusion, however, we ought perhaps exempt the United States, where senior full professors might earn no more than a beginning lawyer or physician.[69]

The professor had no place to rise but into administration or into another professorship. In Germany, where the universities competed fiercely for top men, professors changed institutions (and improved their finances and fringes) with some regularity. We find that about one quarter of German university physicists switched chairs, many of them more than once, and others moved up from a technical institute to a university. In France one physicist in five and in Britain only one in ten left one chair for another, and these shifts usually were returns from the provinces to Paris, or from the colonies to Britain. In the United States, after entering a leading school, few people moved.[70]

[65] Oldrini in U.S., Office of Ed., *Report* (1902), Pt. 1, pp. 777, 780; Enriques, *Scientia, 3* (1908), 141–42.

[66] But not always: cf. the case of Bäcklund over Rydberg, referred to in note 25 to Section A.

[67] For German calls see, e.g., Planck, *Sci. Autobiog.* (1968), p. 21; Röntgen, *Briefe an Zehnder* (1935), pp. 31–37; Hertz, *Erinnerungen* (1927), pp. 135–37, 151–55, 207–08; Sachse, *Althoff* (1928); Lorey, *Nachr. der Giessener Hochschulgesch., 15* (1941), 80–132; Paulsen-Lehmann, *Gesch. gelehrt. Unterr., 2* (1921), 705; Paulsen, *Univ. Study* (1906), pp. 83–85, 177; Mayerhofer, *Clio medica, 2* (1967), 55–62; Turner, *Hist. Stud. Phys. Sci., 3* (1971), 163–74; Forman, *Phys. in Weimar Germany* (1968), pp. 101–21.

[68] Holborn, *Germany, 1840–1945* (1969), p. 388. Cf. Tompert, *Lebensformen* (1969), pp. 26, 37.

[69] *Financial Status* (1907), pp. 23–24. We note the shock of a visiting German professor's wife (Frau Max Weber) at discovering that in an American professor's household "his highly educated wife does the cleaning." Quoted by Tompert, *Lebensformen* (1969), p. 38. Cf. Peixotto, *Getting and Spending* (1927); Henderson and Davie, *Incomes and Living Costs* (1928), pp. 4–13.

[70] Counts from our samples (supra, Table B.2): 6 of 61 British full professors, 7 of 38 French, 20 of 77 German (universities only).

C. LABORATORY EXPENDITURES

We turn now to the expenditures of the laboratories in which academic physics was done. Table C.1, itself summarized in Table I, summarizes our results; the entries are largely guesses, except for the regular annual budgets in the first column.

1. The Regular Institute Budget

In Table C.2 we estimate the *regular* annual expenditure on personnel and matériel in the physics institutes of each of our schools for 1900 (1903 for France and Germany). We divide matériel into "equipment" (apparatus and other laboratory costs) and "maintenance" (expenditures

TABLE C.1

Laboratory Expenditures in 1900[a] (1000's of Marks)

	Regular annual budget	Other funding			
		Extraordinary intramural appropriations	Extra-mural grants	Direct philan-thropy	Total
Austria-Hungary	105	30	2	13	45
British Empire	270	20	5	20	45
U.K.	190	15	5	15	35
Other	80	5	0	5	10
France	250	30	2	8	40
Germany	260	100	5	20	125
T.H.	60	20	0	5	25
Univ.	200	80	5	15	100
Italy	150	20	2	8	30
Netherlands	60	20	5	5	30
Scandinavia	60	20	5	5	30
Switzerland	60	20	5	5	30
United States	390	20	10	20	50

[a]"Regular annual budget" is from Table C.2, using its annual rates of change when necessary to obtain values for 1900; "Extramural grants" come from Table C.4, but increased slightly to allow for likely additional contributions from other bodies. Other numbers, especially those for "direct philanthropy," are guesses. (*Prize* monies listed in Table C.4 are treated as personal income and carried forward immediately to Table I.)

for upkeep of plant, including gas, water, electricity, and insurance). "Wages" are the monies paid to technical assistants (mechanics, machinists), janitors, porters, et al., but not the salaries of teaching assistants, whom we treat as academics. We also give in Table C.2 "official budgets" for 1900 and 1910, which we use to compute a rate of increase in the support of physics. These official budgets are figured very differently from place to place, as appears from the breakdowns in Table C.3. In particular, they sometimes include salaries for teaching assistants, unexpended credits, printing costs, etc., and so exceed the sum of equipment, maintenance, and wages; or, as is more often the case, they omit maintenance or wages, and fall below the sum. Neither the official budget nor the expenditures for matériel and wages indicate actual spending very accurately: for one thing, many institute directors habitually overran their budgets;[1] for another, large amounts, which we discuss in Section C.2, were made available in irregular or extraordinary one-time grants.

Germany. Outlays for all Prussian university institutes were 6.9 million marks in 1900 and 11.0 million in 1910.[2] University physics rose in Germany at the same rate, as Table C.2 shows. Some institute budgets increased much more rapidly than others, however, and great disparity existed between large and small institutes, between the 4,000 marks of Rostock and the 28,000 of Berlin.[3]

We are fortunate in having the matériel budgets for German physics

[1] Lummer at Breslau made it a practice to exceed his allowance; in 1908-09, against a budget of 9,500 marks, he spent 15,000. Breslau, U., *Chronik* (1908-09), p. 43, and earlier years. Cf. Marburg, U., *1527-1927* (1927), pp. 761-63, for overdrafts 1901-06, and Norway, Kirke- og undervisnings-department, *Annaler,* for the same at U. Oslo. In Italy the overruns in the institutes became so frequent that, in 1904, the ministry announced that they would no longer be tolerated. Rome, U., *Annuario* (1904-05), p. 216. See Table C.3, note b, for the overrun by the Paris faculty of sciences in 1900, amounting evidently to 25 percent of its total budget. The discrepancy between budgets and actual expenditures is a systematic error (of defect) in our figures, and if the overrunning of budgets was anywhere a regular and general practice that discrepancy may be an important error.

[2] Prussia, *Statistik, 236* (1914), 16. Budget "für Institute, Sammlungen und den Universitätsgottesdienst," including medical institutes and clinics. Both the number of institutes and the percentage of university expenditures going to them were increasing in 1900. F. Lenz, *Universitätsstat.* (1912), pp. 5, 20.

[3] The median was about 9,500 marks according to figures in *Minerva: Jahrb.* Burchardt concludes that there was a general polarization in university institute budgets before the war, with the rich getting richer. *Konstanzer Blätter, 8* (1970), 71-84. Our data do not support an extension of this conclusion to physics institutes.

TABLE C.2

Regular Annual Physics Laboratory Budgets circa 1900 (1000 Marks)[a]

	"Complete" budget[b]			Official budget		
	Equip-ment	Maintenance	Wages	1900	1910	Δ[c]
Austria-Hungary, 1900:	42	36	33			1.7
Technical Institutes	12	11	10			
Brünn T.H.	1.0	1.0	1.0	1.0	1.0	0
Budapest Polytech.	0.7	1.0	1.0	0.7		
Graz T.H.	1.0	1.0	1.0	1.0	1.9	8
Lemberg T.H.	2.0	1.8	1.5		2.5	
Prague Deutsche T.H.	2.4	2.0	1.5	2.4	2.4	0
Prague Böhm. T.H.	2.5	2.0	1.5	2.5	2.5	0
Vienna T.H.	2.7	2.5	2.0	2.7		
Universities	30	25	23			
Agram U.	1.7	1.2	1.5	1.7	1.7	0
Budapest U.	1.4	1.2	1.5	1.4	4.4	21
Czernowitz U.	1.2	1.0	1.0	1.2	1.7	4
Graz U.	4.5	4.0	4.5	5.7	5.7	0
Innsbruck U.	2.4	2.0	1.5	2.4	2.4	0
Klausenburg U.	2.0	1.8	1.5	2.0	3.0	5
Krakau U.	2.5	2.0	1.5	2.5	2.5	0
Lemberg U.	1.6	1.2	1.5		1.7	
Prague Deutsche U.	3.2	3.0	2.5	3.2	3.2	0
Prague Böhm. U.	2.5	2.0	1.5	2.5	2.5	
Vienna U.	7.0	6.0	5.0	4.0		
British Empire:						
U.K., 1909:[d]	100	100	100			
Aberdeen U.	3.1					
Aberystwyth U.	0.6		2.6			
Bangor U.		5.4	1.6			
Belfast Queen's	3.0					
Birmingham Mason		11.4	7.2			
Bristol U.	1.1		2.5			
Cambr. Cavendish	13.5		12.7			
Cardiff U.	0.8		2.3			
Cork U.						
Dublin Trinity Coll.	2.8					
Dundee U.	3.0		2.4			
Edinburgh U.						

(Footnotes for table on page 64)

TABLE C.2 (*continued*)

	"Complete" budget[b]			Official budget		
	Equip-ment	Main-tenance	Wages	1900	1910	Δ[c]
Galway Queen's						
Glasgow U.	1.3	11.1				
Leeds Yorkshire	1.8		3.0			
Liverpool U.	3.1	2.9	7.6			
London U. Coll.	6.8		6.1			
London King's	1.6		5.1			
London R. Coll. Sci.						
Manchester Owens	8.2	8.6	4.5			
Newcastle Durham	1.4		5.0			
Nottingham U.	2.6		1.7			
Oxford U.	5.5	6.0	11.0			
St. Andrews U.	1.5		4.1			
Sheffield U.	3.6		3.1			
Other, 1900:	26	25	26			
Adelaide U.	1.0	1.0	1.5			
Dalhousie U.	1.0	1.0	1.5			
McGill U.	7.0	7.0	7.0			
Melbourne U.	3.0	3.0	3.0			
New Zealand U.	3.0	3.0	3.0			
Sydney U.	3.0	3.0	3.0			
Toronto U.	7.5	7.2	7.3			
France, 1903:[e]	190		104			4.9
Besançon U.	5.3		3.1			
Bordeaux U.	10.0		7.0		118.6	
Caen U.	5.0		4.3			
Clermont-Ferrand U.	5.2		3.0	29.0	36.3	3
Dijon U.	4.8		3.4	33.7	50.8	5
Grenoble U.	5.2		4.0		34.6	
Lille U.	8.1		6.0	46.6	56.8	2
Lyon U.	11.0		8.7	68.3	74.1	1
Marseille U. Aix-M.	6.6		5.3	51.5	142.9	18
Montpellier U.	7.4		5.2	31.5	72.5	13
Nancy U.	14.0		6.9	87.7	184.2	11
Paris U.	74.0		30.9	348.8	435.0	3
Paris Coll. France	3.0		0.5			

(*Footnotes for table on page 64*)

TABLE C.2 (*continued*)

	"Complete" budget[b]			Official budget		
	Equip-ment	Main-tenance	Wages	1900	1910	Δ^c
Paris École Normale	5.0		0.5			
Paris École Polytech.	5.0		2.0			
Paris Éc. Phys. & Chim.	3.0		0.5			
Poitiers U.	4.2		2.7	30.9	36.3	2
Rennes U.	5.5		4.7		54.8	
Toulouse U.	7.2		5.5	47.8	69.1	5
Germany, 1903:	116	110	80			
Technical Institutes	25	17	31			
Aachen T.H.	1.5	1.0	2.0			
Berlin-Charlott. T.H.	4.0	3.0	5.0			
Braunschweig T.H.	1.2	0.5	1.5			
Darmstadt T.H.	3.0	2.0	3.0			
Dresden T.H.	3.5	2.0	3.5			
Hannover T.H.	2.0	2.0	2.0			
Karlsruhe T.H.	3.0	2.0	6.6			
Munich T.H.	5.0	3.0	6.0			
Stuttgart T.H.	1.6	1.0	1.8			
Universities	91	93	49			6.2
Berlin U.	10.5	11.9	5.9	34.8	40.4	2
Bonn U.	4.0	4.4	2.7	9.4	14.6	6
Breslau U.	4.0	4.7	1.1	12.1	21.5	8
Erlangen U.	3.0	3.9	1.8	7.8	14.6	9
Freiburg-i.-Br. U.	2.0	2.0	1.4	4.0	5.3	3
Giessen U.	4.3	2.3	2.0	6.0	8.0	3
Göttingen U.	4.0	4.1	2.5	11.0	26.0	14
Greifswald U.	2.8	3.0	1.3	9.3	11.2	2
Halle U.	4.0	4.4	2.1	9.9	15.2	5
Heidelberg U.	3.5	2.9	1.0	4.9	6.0	2
Jena U.	3.3	3.3	1.4			
Kiel U.	5.0	4.5	2.1	9.9	16.2	7
Königsberg U.	4.0	4.3	1.7	11.0	15.8	4
Leipzig U.	12.5	13.5	6.5	8.0	41.5	40
Marburg U.	4.0	4.5	2.0	8.5	15.2	8
Munich U.	6.0	6.0	3.6	8.1		
Münster U.	2.0	2.1	1.0	6.3	10.6	7
Rostock U.	2.0	1.8	0.3	6.7	9.6	4

(Footnotes for table on page 64)

TABLE C.2 (*continued*)

	"Complete" budget[b]			Official budget		
	Equip-Ment	Main-tenance	Wages	1900	1910	Δ[c]
Strassburg U.	3.0	3.0	4.9	13.4	15.9	2
Tübingen U.	3.4	3.0	1.4	10.1	12.9	3
Würzburg U.	3.5	3.5	2.1		13.2	
Italy, 1900:	57	34	60			3.2
Bologna U.	4.7	2.7	4.3	9.0	15.6	7
Cagliari U.	1.3	1.0	2.0	3.3	2.7	-2
Catania U.	2.3	1.3	2.9	5.2	5.6	1
Genoa U.	2.3	1.3	2.9	5.2	5.6	1
Messina U.	1.8	1.0	2.9	4.7	6.6	4
Modena U.	2.3	1.3	2.9	5.2	5.9	1
Naples U.	5.9	3.4	4.9	10.8	12.6	2
Padua U.	4.7	2.7	5.5	10.2	11.1	1
Palermo U.	3.5	2.0	4.0	7.5	8.5	1
Parma U.	1.8	1.3	4.0	5.8	8.1	4
Pavia U.	4.7	2.7	3.9	8.6	8.5	-0.1
Pisa U.	4.7	2.7	4.3	9.0	20.0	12
Rome U.	9.4	5.4	7.3	16.7	22.0	3
Sassari U.	1.3	1.0	0.8	2.1	2.5	2
Siena U.	2.9	1.7	4.0	6.9	6.9	0
Turin U.	3.5	2.0	2.9	6.4	11.7	8
Netherlands, 1900:[f]	21	18	16			(3)
Amsterdam U.	5	5	3	5.4	20.5	
Groningen U.	4	3	3	6.8	7.7	
Leiden U.	7	5	7	12	16	
Utrecht U.	5	5	3	16	9.4	
Scandinavia, 1900:[g]	23	23	12			7.5
Copenhagen T.H. } Copenhagen U. }	(6.5)	(6.5)	(5.5)	(6.5)		
Oslo U.	5.0	5.0	1.6	4.1	6.6	6
Lund U.	3.2	3.2	1.6	3.2	4.4	4
Stockholm Högsk.	1.8	1.8		1.8	2.7	3
Uppsala U.	4.0	4.0	1.6	4.0	8.2	11
Helsinki U.	2.4	2.4	(1.6)	2.4	5.1	11

(*Footnotes for table on page 64*)

TABLE C.2 (*continued*)

	"Complete" budget[b]			Official budget		
	Equip-Ment	Main-tenance	Wages	1900	1910	Δ[c]
Switzerland, 1900:[h]	28	(15)	(15)			2
Basel U.	2.5		1.0	2.5	2.5	0
Bern U.	3.5			3.5	3.5	0
Fribourg U.	4.5					
Geneva U.	2.5	1.2		2.5	3.3	3
Lausanne U.						
Neuchâtel U.	1.5			1.5	1.5	0
Zürich ETH	9.3	1.9	5.8			(3)
Zürich U.	2.7	1.3	1.3	4.0	5.1	3
United States, 1900:[i]	136	140	86			(15)
California U.	6.3	6.3	7.5	13.8		
Chicago U.	10.0	10.0	6.0			
Clark U.	2.5	2.5	6.0	11.0		
Columbia U.	6.3	8.0	0	6.3	12.9	10
Cornell U.	11.9	10.0	6.0	27.9		
Harvard U.	20.0	20.0	6.3			
Johns Hopkins U.	8.0	8.0	6.0			
Illinois U.	4.0	4.0	3.5			
Mass. Inst. Tech.	11.0	11.0	6.0	28.0		
Michigan U.	5.1	5.0	3.3	5.1	14.2	18
Minnesota U.	7.2	7.0	1.0	7.2	19.8	18
Missouri U.	1.5	2.0	3.0			
Nebraska U.	11.5	11.5	3.2	26.2		
New York U.	4.0	4.0	3.0			
Northwestern U.	1.9	2.5	2.0			
Ohio State U.	0.8	2.0	0			
Pennsylvania U.	4.0	4.0	3.0			
Princeton U.	2.1	4.0	3.0			
Stanford U.	4.0	4.0	3.0			
Wisconsin U.	8.3	8.0	6.0			
Yale U.	5.7	6.0	6.0	5.7	14.7	17

(*Footnotes for table on page 64*)

FOOTNOTES FOR TABLE C.2

[a]See text and notes thereto for explanation of categories, procedures, and sources.

[b]Where the "complete budget" (sum of "equipment," "maintenance," and "wages") differs significantly from the "official budget"—as it does in the overwhelming majority of cases—the "complete budget" is our estimate, as is a fortiori the distribution over the three categories of expenditure. These estimates, being essentially guesses—informed and considered guesses—ought according to our conventions appear within parentheses. The year to which these estimates refer is 1900-01 for all countries except the United Kingdom, France, and Germany. For the United Kingdom the estimates are for 1909-10; for France and Germany, 1903-04.

[c]Rate of change in official budget per annum as percent of 1900 level.

[d]The figures entered in the "equipment," "maintenance," and "wages" columns, or a pair of these columns, are the documentarily attested categories of expenditure for 1909-10 or a neighboring year. The British data, apart from Oxbridge, are from G.B., Bd. of Ed., *Reports*, the Scotch from the "Annual Accounts" and "Statistical Accounts" as given in the bibliography, the Welsh in part from the Committee on the Welsh University. The figures for Cambridge and Oxford were kindly supplied by Peter Heimann and Alan Chapman, respectively. In view of our ultimate goal, a figure for the year 1900, we have not included Reading and Southampton which became University Colleges on the government grant in 1902. Their inclusion would have added perhaps 15,000 marks to the total for 1909-10.

[e]Note that the "official budget" in France is for the *entire* faculty of science.

[f]The Dutch "official budgets" are taken from *Minerva: Jahrb.*.

[g]See notes to Table A.4 for sources for all Scandinavian institutions but Stockholm Högskola, for which we use the budgets in *Minerva: Jahrb*. The official budgets are for equipment only; in 1910 wages at Lund and Uppsala were, respectively, 3,300 and 4,000 marks. The equipment budget for the Copenhagen T.H. (which also housed the University's physics laboratory) is taken arbitrarily as twice the average of the equipment budgets of the other schools. (The ratio of wages would suggest a factor of four, which we reduce to account for the share of applied physics.) Maintenance is taken equal to the equipment expenditure, the ratio obtaining in Germany.

[h]*Switzerland:* Basel, U.: Basel (Canton), *Staats-Rechnung* (1900), p. 13; (1910), p. 15. Bern, U.: Bern (Canton), *Staats-Rechnung* (1900), p. 20; (1910), p. 26. Freiburg, Schw., U.: "Jahres-Etat 5000 Franken für Neuanschaffung von Instrumenten, 500 Franken für die Bibliothek," *Bericht* (1899-1900), p. 30. Geneva, U.: Geneva (Canton), *Budget* (1900), pp. 50-51; (1910), p. 58. Neuchâtel, U.: Neuchâtel (Canton), *Décret portant budget de l'état* (1900), pp. 50-51; (1910), p. 58. Zürich, ETH: projected from 1910 budget on the assumption of 30 percent increase in the ten years 1900-1910: Zürich, ETH, personal communication, 20 Mar. 1973; H.F. Weber, "Bericht" (1910); P. Weiss, "Bericht" (1910). Zürich, U.: personal communication, 19 Apr. 1973; Zürich (Canton), *Staatsrechnung* (1900), p. 88; W. Meyer, *Finanzgesch.* (1940), pp. 105-06, Anhang, Table 6.

[i]Only where an entry appears in the "official budget" column do we have a documentary basis for our estimates for the institution in question. These figures, which are a medley of data for 1899 and 1900, are drawn from the histories of the physics departments of California, Cornell, and Minnesota, from annual reports as listed in the bibliography and notes to Table A.4, and from Cooke, *Acad. Efficiency* (1910).

institutes in 1903. These figures were collected in 1904 by Ferdinand Lot for comparison with French expenditures.[4] The breakdown into equipment and maintenance, however, is an estimate, guided by data from institutes whose budgets we know in detail.[5] We usually must estimate how much of the difference between the matériel budget and the official budget is due to wages, and how much to teaching assistants' salaries. For each university the total expenditures refer not only to experimental physics institutes but also to mathematical physics institutes and seminars, although these were sometimes more concerned with mathematics than with physics. We estimate that matériel expenditures for mathematical physics (primarily for books and journals, demonstration models, and perhaps a share of maintenance) totaled some 5 percent of all German university physics matériel expenditures in 1903, and roughly the same proportion of wages.

For the technical institutes we have less data, and for about half of all schools we have estimated physics funds by comparing with schools where these are known and prorating by the numbers of students and total receipts.[6] We assume that the ratio of maintenance to equipment expendi-

[4] Lot, *Situation faite.*

[5] And yet we have probably credited too much to equipment: cf. the insufficiency of the sizeable sum (3,900 marks) budgeted for maintenance at U. Breslau (*Chronik* [1904–05], p. 38), and the complaints of Kayser (Bonn, U., *Chronik* [1908], p. 46): "In the early years [of the century] virtually no new apparatus could be bought since, owing to the great number of men in the practical courses, the consumption of gas, electricity, glass, chemicals, photographic plates, etc., has eaten up the entire budget." Our sources:

Strassburg had 13,445 marks total, of which 2,625 were for two teaching assistants and 4,820 for the 4.5 employees, leaving 6,000 for the rest. Alsace-Lorraine, *Landeshaushaltsetat* (1900), pp. 108–09. For Leipzig: Wiener and Des Coudres in Leipzig, U., *Festschrift, 4,* pt. 2 (1909), 56–69. For Tübingen: Württemberg, Kammer der Abg., *Verhandl.,* 34. Landtag (1899), Beilagenband 1, Heft 6, 238–39; *ibid.,* 36. Landtag (1904–05), Beilb. 2, Heft 6, 240–41. For Munich: Bavaria, Kammer der Abg., *Verhandl.* (1899–1900), Beilb. 4, 315. In all cases we have separated assistants' salaries, estimating where necessary. That wages and salaries together took up 40 percent of the sum of Prussian physics institute budgets about 1910 can be inferred from the figures in *Minerva: Jahrb.,* for 1910–11 and 1911–12, since after 1910 Prussia removed these sums from the total accounted for under institutes. Prussia, *Statistik, 236* (1914), 9, n. 2. We have allowed for a 6 percent average increase in overall budgets.

[6] For general expenses of the Technischen Hochschulen, see Prussia, *Stat. Jahrb., 1* (1904), 170–71; Lexis, *Unterrichtswesen, 4,* pt. 1 (1904); Baden, Stände-Versamml., and 2. Kammer, *Verhl.,* Landtag 1899–1900, Beilb. 3, Hauptabt. 3, 184–85. For Stuttgart, see Württemberg, Kammer der Abg., *Verhandl.,* 37. Landtag (1907–09), Beilb. 5, Heft 6, 108–11; for Karlsruhe, where, in 1900, physics received 4,000 marks

ture on physics was somewhat smaller in the Technischen Hochschulen, where several departments shared one building, than in the less efficient separate university institutes.

France. Following the disaster of 1870, France overhauled her educational system, and particularly instruction in science. Between 1877 and 1887 the matériel expenditures for all science increased from 350,000 to 570,000 marks. Thereafter it remained flat until just before 1900, when shifts in organization and accounting—the transference of basic science courses for medical students to the science faculty in 1893 and, after 1896, the retention by the universities of student fees, formerly reserved for the state—almost doubled the total, which reached 970,000 marks in 1903.[7] But state support as a whole (including personnel) increased only some 10 percent between 1887 and 1900, while the matériel funds provided by the state to all faculties *decreased* 30 percent.[8] Between 1900 and 1910 there was a further decrease of 6 percent.[9] The Paris faculty of sciences received some 120,000 marks in student fees in 1901 and 250,000 in 1911, but because state funding failed to increase, the science matériel budget (given in Table C.3 for 1899) rose much more slowly, at about 2 or 3 percent per year.[10]

We have not found any information on the funds spent separately on physics in France. We give as "official" budgets those of science faculties

---

plus 3,500 for two teaching assistants and around 6,500 for four employees, see Lehmann, *Phys. Inst. Karlsruhe* (1911), pp. 73–75, 84–87. The firm data for T.H.'s in Table C.2 are totals of wages and equipment in Aachen and Braunschweig; wages in Braunschweig, Karlsruhe, and Stuttgart; the total excluding wages for Stuttgart (4,000 marks); assistants' salaries (3,500) and the total of equipment plus some maintenance for Karlsruhe (4,000); and the total for Dresden (11,800 including assistants' salaries).

[7]Compiled from France, Min. instr. publ., *Statistique, 3* (1889), and Lot, *Situation faite* (1905). From 1897 a student in Paris paid 240 to 970 marks a year for laboratory fees (all to the university), 360 to 680 marks for the license (160 to the state), and 170 marks for the doctorate (all to the state). There were also registration fees. France, Min. instr., publ., *Recueil des lois, 5* (1898), 769–70.

[8]Typically increments in one budget category were compensated by decreases in others; France, Min. instr. publ., *Statistique 4* (1900), 462-67, 484-99.

[9]France, *Journal officiel* (1887), p. 991; (1900), p. 2319; (1910), p. 3176. The 8 percent inflation of the franc from 1890 to 1910 caused further loss. France, Inst. Nat. Stat., *Annuaire stat., 58* (1951), 515*.

[10]Liard, *Univ. de Paris* (1909), pp. 84–85; Paris, U., Conseil, Conseil acad., *Rapports* (1901–02), p. 77; Paris, U., Conseil, *Rapport* (1910–11), p. 57. About one quarter of the fees collected in 1910 were for laboratory use. Budget compiled from France, Min. instr. publ., *Statistique, 4* (1900), eliminating salaries.

as a whole,[11] and in the left side of Table C.2 credit physics with one sixth of the total science matériel and wages.[12] This is certainly a conservative estimate since faculty, assistants, and laboratory area[13] for physics were each about one fifth of the respective totals for the science faculties, and in both British and German universities physics took between one fourth and one fifth of the total science matériel expenditures.[14] We also add estimates of expenditures for our French schools outside the university system.[15]

The funds spent on physics in France were substantial. The Sorbonne sheltered three independent laboratories for physics plus two for teaching physics to medical students, and the most important of all, Lippmann's advanced research laboratory for graduate students and professors, drew a budget of 10,000 marks a year.[16] The polarization between large and

[11] In general from *Minerva: Jahrb.* For Paris for 1910 we use the 1909 value in Liard, *Univ. de Paris* (1909), pp. 84-85. For Bordeaux, Dijon, Marseille, Montpellier, and Rennes, 1905 science faculty incomes are given in *Minerva: Jahrb.* (1907-08), p. 922. Altogether, we have culled from *Minerva* both 1905 income and 1910 expenditures for eight French universities; the latter average 0.985 times the former (faculties sometimes reserved part of their income for extraordinary outlays, hence the apparent decline). We estimate 1910 science expenditures for the five above-named universities by multiplying the 1905 incomes by 0.985.

[12] Wages from France, Min. instr. publ., *Statistique, 4* (1900), and matériel from Lot, *Situation faite* (1905), giving figures for 1903. We have added 2,500 marks to the physics matériel of the U. Montpellier for a special grant. France, Ministère de l'instruction publique, "État comparatif par facultés," Archives Nationales (A N F[17] 13396, budget for 1903); Montpellier, *Livret* (1902), p. 124.

[13] France, Min. instr. publ., *Statistique, 3* (1889).

[14] In Germany physics took about 18 percent of institute expenditures, not including medical institutes. *Minerva: Jahrb.* (1899-1900). In Britain, physics got 22 percent of "Departmental Maintenance and Lab. Expense" of the science departments and a bit more of wages. G. B., Board of Ed., *Reports* (1909-10).

[15] Some figures for the École normale and the Collège de France are in France, Min. instr. publ., *Statistique, 4* (1900). We add 10,000 marks to the funds of the U. Paris and a problematic 3,000 to those of the École normale as the contributions of the École pratique des hautes études.

[16] Berget, *La Nature, 26* (1898), 225-27, including lighting and heating and possibly some wages. *Nature, 58* (1898), 12-13. Cf. H. Crew, "Diary," 25-26 July 1895: "University of Paris—Sorbonne—Faculty of Sciences—is very completely equipped in Chemistry, Physics, Physiology & Geology & Botany. In Physics they have 4 men—Bouty, & Lippmann & Pellat & Leduc—besides endless assistants. . . . Lippmann says they have spent 30,000 francs for apparatus during the last two years—building cost several million *dollars*. Perhaps the method of laboratory training here employed—the student apparently making no apparatus—accounts for the fact that the young men of France are, so far as I can see, doing very little in physical research."

small universities was particularly marked in France. Only Nancy could compete with Paris in science: the other thirteen universities had a median matériel budget for all sciences of about 30,000 marks in 1903; the comparable figure for the eighteen smaller German universities is 45,000 marks.[17]

United Kingdom. The government grant to universities and university colleges, 0.6 million marks in 1890, increased to 1.6 million in 1900 and 4.3 million in 1910.[18] These figures are still very low compared with the 16.2 million marks the French government spent on her universities in 1910[19] or the 16.8 million Prussia alone spent in the same year.[20] But private endowment in Britain was greater than on the continent and went far to make up for the state's stinginess. The annual expenditure of our thirteen English and Welsh redbrick universities rose from 6.5 million marks in 1899–1900 to 11.2 million marks in 1910–11.[21] The consequences for physics in the United Kingdom may be gathered from Table C.2. (Again note the great range in budgets.) These data refer to 1909–10, for which our principal source, the *Reports* of the Board of Education, gives more complete data than for any previous year.[22] Our scanty data for Scotland and Ireland are no better for 1909–10, but are taken from it for the sake of comparability.[23]

[17]Compiled from Lot, *Situation faite* (1905), pp. 213, 217. The German universities not included in this calculation are Berlin, 324,000 marks, Leipzig, 113,000, and Munich, 116,000. For complaints from provincial universities, see France, Min. instr. publ., *Enquêtes*, Nr. 101 (1909–10), pp. 70–71 (Aix-Marseille), 109 (Bordeaux), 135 (Clermont).

[18]MacLeod and Andrews, "Selected Science Statistics" (1967), p. 15. For a detailed breakdown for England and Wales alone see G.B., Board of Ed., *Reports*, (1910–11), pp. xxviii–xxix.

[19]Including the École normale, Collège de France, École pratique des hautes études, and Muséum d'histoire naturelle. France, *Journal officiel* (1910), p. 3176.

[20]State funding only, ordinary plus extraordinary. Prussia, *Statistik, 236* (1914), 16, 17.

[21]G.B., Bd. of Ed., *Reports* (1910–11), p. xviii; (1901), passim.

[22]G.B., Bd. of Ed., *Reports* (1909–10) usually list both "departmental maintenance and laboratory expense" and "departmental wages." The former does not, in the great majority of cases, include "maintenance" as understood by us, but rather corresponds to our category "equipment." We have thus been forced to estimate the expenditure in our category "maintenance," and take it to be about 5 percent of overall university maintenance. Where no separate figure for physics department wages is given in the *Reports* we prorate total science wages at 30 percent, the average for schools where the breakdown is given.

[23]For Aberdeen, Glasgow, and St. Andrews we draw upon the "Annual Accounts" and "Statistical Accounts" listed in the bibliography. For Dublin and Belfast we take figures from G.B., Parl., H. of C., *Sess. Papers*, 1911, *59*, 388–96.

The total regular laboratory expenditure thus estimated has been projected back to 1900 on the precarious assumption that physics funds increased at the same rate as redbrick university funds in general.[24] Outside the United Kingdom we have referred our estimates to the year 1900.[25]

Table C.3 gives the breakdown of the budget of 1904 of the physics institute completed in Manchester in 1900.[26] As in Germany, the institute's maintenance took up about half its matériel budget. We also give the Cavendish budget,[27] which evidently excludes some maintenance items (repairs, insurance, janitorial expenses) doubtless defrayed from general university funds.

Once again student fees contributed substantially to the budget. For example, in Manchester in 1900 students paid nearly 270 marks per session (30 weeks) for a laboratory course meeting three days a week.[28] Their fees made up 45 percent of the college's income that year, as against 9 percent from the British government.[29] So in Britain, as on the continent, the growing student body directly increased the funding for physics.

United States. The total annual income of our twenty-one leading American universities in 1907 was some 70 million marks, exclusive of gifts; they taught some 43,000 students. At this time the German universities also taught over 40,000 students, but had an income, excluding gifts, of only some 40 million marks.[30] Even allowing for different ac-

[24]The calculation agrees fairly well with one made by taking physics to have 8 percent of all laboratory equipment funds at large schools and 12 percent at small schools (those spending less than £20,000 per year), which are the 1909 proportions.

[25]Outside the U.K. we have the Toronto budget in detail from Toronto, U., *President's Report* (1902), p. 32, and more general information in Montreal, McGill U., *Annual Report* (various years) and Eve, *Nature, 74* (1908), 272-75.

[26]G. B., Board of Ed., *Reports* (1904-05), p. 54. Cf. Manchester, U., *Phys. Lab.* (1906), pp. 127-29.

[27]Kindly furnished by Peter Heimann from Cambridge U., *Reporter*, 18 March 1901; our 1910 Cavendish budget is from *ibid.*, 20 Mar. 1911. Cf. Thomson, *Recollections* (1937), pp. 124-26, and in Brit. Ass. Adv. Sci., *Adv. of Sci.* (1931), p. 12.

[28]Also some 70 marks for three lectures a week. G. B., Board of Ed., *Reports* (1900-01), pp. 281-82. Student fees provided 85 percent of the Cavendish budget given in Table C.3. Cambridge U., *Reporter*, 18 March 1901. Fees contributed 33 percent of the income of all university colleges in England in 1909. G. B., Board of Ed., *Reports* (1909-10), p. xx. At Edinburgh in 1909 the proportion was 50 percent. G. B., Parl., H. of C., *Sess. Papers*, 1911, *60*, 63.

[29]Compiled from G. B., Board of Ed., *Reports* (1900-01), pp. 324-25.

[30]Budgets are for 1906, from Prussia, *Statistik, 236* (1914), 16, and *Minerva: Jahrb.* (1907-08), with appropriate estimates; students are for 1905, from Prussia, *Statistik, 204* (1908), 29. See U.S., Office of Ed., *Report* (1900-01), *1*, 797-810.

TABLE C.3

Representative Budgets of Leading Institutions (Marks)

		Percent of matériel budget
University of Berlin Physics Institute, 1909[a]		
Wages	9,930	
Equipment		
Instruments and repairs	4,000	20
Laboratory needs [*Laboratoriumsbedarf*]	5,000	25
Workshop materials	1,000	5
Maintenance		
Heating material	4,000	20
Gas, water, and electricity	3,500	17
Cleaning and fees [*Abgaben*]	2,760	14
Contingency	2,214	
	32,404	100
University of Paris Faculty of Sciences, 1899[b]		
Wages	185,000	
Equipment		
Expenses of courses and laboratories	120,000	41
Apparatus and collections	26,700	9
Expenses of laboratory exercises and exams.	77,000	26
Maintenance		
Heat and light	48,000	16
Upkeep of buildings	14,200	5
Upkeep of furniture	3,300	1
Miscellaneous	5,800	2
	480,000	100
Cambridge University, Cavendish Laboratory, 1900[c]		
Wages	10,250	
Professor's share of fees, to assistant	2,800	
Equipment	7,180	76
Maintenance		
Coal	1,200	13
Gas, water	720	8
Cleaning	250	3
	22,400	100

TABLE C.3 (*continued*)

		Percent of matériel budget
University of Manchester Physical Laboratories, 1904[d]		
Wages	4,100	
Equipment		
Apparatus	3,900	28
Current expenses	3,200	23
Maintenance		
Fittings and repairs	500	4
Coal	900	6
Gas, water, and electricity	3,300	24
Cleaning, insurance, and fees	2,100	15
	18,000	100
Harvard University, Jefferson Physical Laboratory, 1904[e]		
Wages and services	5,650	
Equipment		
Collections and supplies	4,150	24
Research funds	1,850	10
Maintenance		
Fuel	2,200	13
Water, lighting, and electricity	4,100	23
Repairs and improvements	500	3
Care and cleaning	4,300	24
Telephone and insurance	550	3
	23,300	100

[a]Rubens in Lenz, *Gesch. U. Berlin, 3* (1910), 296.

[b]France, Min. instr. publ., *Statistique, 4* (1900). The titles which we group under "equipment" include some maintenance. The de facto matériel budget given in "Séances, Conseil de l'Université de Paris, 18 June 1900, Annexe: Faculté des Sciences, Compte de l'Exercice 1899, dépense réellement faite" (AN: AJ[16] 2575) is 25 percent larger but with only one significant difference in the categories and distribution: 14,600 marks introduced under the heading "Emploi des revenues des dons et legs."

[c]Cambridge U., *Reporter,* 18 March 1901, courtesy of P. Heimann. The budget is probably not complete.

[d]G.B., Board of Ed., *Reports* (1904-05), p. 54.

[e]Harvard U., *Treasurer's Report* (1904-05). The budget is probably not complete.

counting methods, for the doubling of German salaries by student fees paid to the professor, and for the difference in cost of living, it is clear that American universities were not behind other countries, especially when one considers that gifts to universities were greater in the United States than elsewhere.[31]

We cannot say exactly how much of the wealth of American universities found its way into physics, but some physics departments enjoyed budgets on a European scale. In Table C.3 we give the budget of the Harvard physics laboratory in 1904, which compares favorably with that of a European institute.[32] As an example of the rapid changes that were under way, we may take the physics department of the University of Minnesota. Around 1900 it paid 2,400 marks for apparatus and supplies, and 1,000 for a mechanic; the university contributed to maintenance. By 1913, having built a new laboratory and doubled the size of its teaching staff, it was spending 16,000 marks annually on apparatus and supplies and 9,500 on wages. Like Harvard, it derived some of its income from endowed research funds.[33]

Our totals for the United States are less accurate than the totals for other countries because no central information collecting agency existed and because the budgets follow differing accounting systems. In most cases we have had to estimate maintenance and wages, using as a guide figures for universities whose arrangements we know.

Other Countries. In Austria-Hungary we have proceeded much as in Germany, although with less data; in general we have equipment figures for universities and for some Technische Hochschulen, and estimate other numbers guided by data from comparable German cases.[34] It appears that,

[31] Data on incomes of individual universities in *Minerva: Jahrb.*

[32] Harvard U., *Treasurer's Report* (1903); cf. Goldberg, "Physics at Harvard," (1962) (AIP). Cooke, *Acad. Efficiency* (1910), p. 60, reports that in 1907 some 20,000 marks in student fees were received and spent by the physics department in addition to this official budget. We note that the figures in Cooke's category "direct expenses," less salaries, for Harvard in 1907 (and for Columbia, MIT, Princeton, and U. Toronto), considerably exceed our figures (or projections) for matériel plus wages for that year. Cooke, *Acad. Efficiency* (1910), pp. 100, 121–23. We take it that at least part of the discrepancy lies in the accounting: Cooke's category probably includes monies we have placed in "extraordinary intramural appropriations" and "extramural grants" (Table C.1). In drawing up Tables C.2 and C.3, we have used Cooke's figures only for their indication of the relative share of the several budget categories. Cooke also gives, as indirect costs, physics' share of the general operating expenses of the university, which we likewise ignore.

[33] Erikson, "U. Minnesota Dept. of Phys.," pp. 14–17, 37, 266.

[34] Equipment figures for 1900 and 1910, given as "official budget," are from *Minerva: Jahrb.* Cf. Hellmer in Brünn, T.H., *Festschr.* (1899), pp. 62, 75.

in the provinces at least, funding in 1900 remained at levels established twenty years earlier.[35] We have more information for Italy, where we estimate wages from the numbers of employees and their approximate average wages.[36] We assign a maintenance expenditure equal to the official equipment budget, which was the situation in most countries.[37] Table C.2 also presents some estimates for regular annual budget in Switzerland, the Netherlands, and Scandinavia.[38]

2. Special Grants

The monies so far discussed constitute the ordinary budget of the physics institute. Table C.1 also estimates the amount of extraordinary funds flowing to the laboratories from various public and private sources. Although these funds came irregularly, they were of the first importance in supporting research activity.

a. Extraordinary Intramural Appropriations

In Germany extraordinary or one-time appropriations by or on the motion of the education ministries were especially important, running about one third of the total university matériel budget.[39] Although such

[35]The equipment allowance at U. Czernowitz, e.g., set at 1,300 marks in 1881, had decreased to 1,200 by 1899, the difference going into the annual budget (250 marks) attached to the theoretical chair. Czernowitz, U., *Festschr.* (1900), p. 123. Cf. the situation in chemistry at Prague, in Prague, U., *Dtsch. K.-F.-Univ.* (1899), pp. 14–17.

[36]The equipment budget comes from *Minerva: Jahrb.;* wages are from data on employees in Italy, Min. pubbl. istr., *Annuario* (1900–01), and the following average salaries: meccanico, 1,800 marks; asst. mecc., 1,200; custode, 1,350; inserviente, 1,100 in large and 850 in small institutes. These numbers are straight-line averages of the de facto wages of 1881 and the official beginning wages of the law of 1909. Italy, Min. pubbl. istr., *Bolletino,* 7 (1881), 31–120; Pisa, U., *Annuario* (1909–10) pp. 277–336. Under "official budget" we give the sum of equipment and wages; for 1910 wages we use the staffs given in the Ministero's *Annuario* and the 1909 wages. For Bologna, Messina, and Naples we lack 1910 equipment figures; for Messina we use that of 1905 (a nice increase over the 1900 figure), and for Bologna and Naples we use the 1900 figures plus 10 percent.

[37]At Turin heat and light added 55 percent to the equipment budget; the rest would be accounted for by cleaning, repair, insurance, water, etc. Brunhes, *RIE, 41* (1901), 400–01.

[38]Our data for Japan and Russia suggest only the generalizations that physics increased rapidly in the former and stagnated in the latter. Watanabe, *Congrès internat. hist. sci.* 10th (1962), *Actes, 1,* 197–208; Shizume, *ibid.,* 208–210; Joffe, *Uspekhi fizicheskikh nauk, 33* (1947), 453–68; Vavilov, *ibid., 28* (1946), 1–50; Kydriavtsev, et al., in Figurovskii, ed., *Istoriia Estestvoznaniia v Rossii, 2* (1960), 318–501; Lazarev, *Ocherki Istorii Russkoi Nauk* (1950), pp. 66–74.

[39]Average for Prussia, 1895–1900, calculated from Prussia, *Statistik, 167* (1901), 12, 13. By "matériel" we mean total funds for "Institute, Sammlungen und den Universitätsgottesdienst." Among the special grants listed in *Hochschul-Nachrichten*

funds might be granted where and when the need arose, more often they served as counters in the competition between universities for outstanding men. A German state might build a new wing or an entire new institute to attract or keep a good professor; at the very least it would grant some thousands of marks to improve the lighting in the lecture hall or the collection of apparatus. Between 1888 and 1906, when its head was changed four times, the physics institute of the University of Berlin enjoyed special grants totaling over 100,000 marks, although no new construction was undertaken.[40]

The French government gave few special grants of consequence around or after 1900. In 1901, for example, the Paris science faculty received a total of some 25,000 marks in extraordinary appropriations,[41] a sum less than 7 percent of the science matériel budget; other schools probably got no more in proportion to their size.[42]

In the Anglo-Saxon countries, where universities were not national institutions, the central government gave few special grants for physics. Special grants from the universities, often derived from the monies provided by local governments to which we earlier referred, could, however,

---

for the two years 1899-1900, over half a million marks, plus unspecified further sums, went to science. Some sample special grants to physics institutes: U. Bonn, 1901-10, 22,000 marks chiefly for apparatus (*Chronik*); U. Halle, 1901-02, 5,000 for apparatus, 2,000 for wiring (*Chronik,* p. 43); 1902-03, 6,000 for electric motor (*Chronik,* p. 44); T.H. Karlsruhe, 1889-1901, 77,000, of which 30,000 in 1892 for a machine room (Lehmann, *Phys. Inst. Karlsruhe,* 1911, pp. 73-75); U. Königsberg, 1901-02, 5,650 primarily for instruments (*Chronik,* p. 36); U. Marburg, 1901-06, 35,000 for special equipment (Marburg, U., *1527-1927* [1927], pp. 761-63, and *Chronik* [1910], p. 47); U. Strassburg, 1898-99, 14,000 for renovation (Alsace-Lorraine, Landesaussch., *Etat* [1898-99], pp. 86-87).

40 Rubens in Lenz, *Gesch. U. Berlin, 3* (1910), 288-95. The grants were used to buy apparatus, a bank of batteries, shutters to darken the lecture hall, etc. Lummer requested 60,000 marks for improvements when taking over at Breslau; he got 40,000 (*Chronik* [1905-06], p. 40).

41 Paris, U., Conseil acad., *Rapports* (1900-01), p. 74. The substantial extraordinary credits listed in France, Min. instr. publ., *Statistique, 3* (1889), 650-53, went mainly for scholarships and examinations.

42 Lot (*Situation faite,* [1905], pp. 218-20) found the total *regular* budgets of the French faculties of science to be "almost sufficient," and not shameful compared to German expenditures (1 million marks against 1.5 million, about in proportion to the populations); "rather the affliction of our institutions, especially in the provinces, is that the state has not provided suitable quarters or expended once and for all the sums necessary to complete the apparatus and collections," i.e., has not given enough *crédits extraordinaires.* On the plus side we can point to 4,900 marks for refurbishing M. Curie's laboratory. Paris, U., Conseil de l'U., Conseil acad., *Rapports* (1908-09), p. 31.

be considerable. In the United Kingdom we give extraordinary grants for 1900 in the same proportion to equipment expenditures as prevailed in 1909–11;[43] for other countries we offer only informed estimates.[44]

b. Extramural Grants and Prizes

In each leading country there were organizations giving grants and prizes specifically for pure research. Their annual beneficence averaged for the years 1896–1905 appears in Table C.4. It may not seem a large addition to the total funding of physics. In fact these grants played an important part, for they were almost the only substantial sums of money that organized bodies gave explicitly for pure research in physics at the turn of the century.

To emphasize their importance, we enter in Table C.4, column 2, the proportion of extramural grants and prizes awarded for work in perhaps the most rapidly growing branch of fundamental physics in this period: the study of cathode rays, X rays, and radioactivity.[45] About one third of these grants went for experimental work in ray physics in France, Germany, and Austria. In Britain the proportion was far smaller. Although the Royal Society did not altogether ignore modern physics, giving in particular over 6,000 marks to Oliver Lodge for his ether drag experiment, this sum compared very unfavorably with the continental academies' support for up-to-date research.

The Royal Society funds came chiefly from a government grant of 83,-000 marks a year, which had to support not only all the sciences but some technical and medical research as well, and this grant remained at the same level from 1876 to 1914.[46] The continental academies relied less on gov-

[43] G. B., Board of Ed., *Reports* (1909–10), pp. xx–xxi; (1910–11), p. xvii. We identify as extraordinary grants the sums quoted under "capital expenditures"; in 1909–11 these added 23 percent to those quoted under "departmental maintenance and lab. expenditures."

[44] In the decade 1900–10 U. Lund received 11,000 marks and T.H. Copenhagen 155,000 for refurbishing and restocking their physics laboratories. Leide, *Fysiska institutionen* (1968), pp. 131–32; Copenhagen, Polytekniske Laereanstalt, *Program* (1910), pp. 9–13; Lundbye, *Den polytekn. laereanst.* (1929), pp. 363–68. We arbitrarily increase the sum to 200,000 and enter the average per year for Scandinavia in Table C.1.

[45] That this was a recognized branch of physics at the time appears from the session "Magnéto-optique, rayons cathodiques, uraniques, etc." at the Congrès international de physique, Paris, 1900.

[46] For details of the administration of the grant see MacLeod, *Hist. Journal, 14* (1971), 323–58; also MacLeod, *Minerva, 9* (1971), 197–230; for medals, MacLeod, *Notes and Records, 26* (1971), 81–105. According to the Royal Society's *Yearbook,*

TABLE C.4

Grants and Prizes of Leading Academies circa 1900

	To and for all basic physical research		"Ray" physics as percent of all basic physics[c]	Nobel Prizes[d]	
	1000's of marks per annum[a]	Δ, percent per annum[b]		Number, 1901-14	1000's of marks per annum
Austria-Hungary	2.9	28	35	0	
Grants	2.0	19	25		
Prizes	0.9	97	60		
Belgium	0	0	0	0	0
British Empire	4.7	-6.5	1.5	2	24.2
Grants	4.4	-7.0	0.9		
Prizes	0.3	0	50		
France	17.0[e]	13	20	2	24.2
Grants	1.3		50		
Prizes	15.7	7.2	17		
Germany	4.7	16		4.5	54.4
Berlin Acad.	3.1		33		
Grants	2.6	3.0	20		
Prizes	0.5	0	100		
Göttingen Acad.	1.6				
Grants	1.6	17	27		
Prizes	0	0			
Italy	9.3	30	20, 40[g]	0.5	6.1
Japan	0	0		0	0
Netherlands				3	36.4
Russia				0	0
Scandinavia				0	0
Switzerland				1	12.1
United States[f]	(10)	(15)	(0)	1	12.1

[a]Averages of actual awards in the period 1896-1905; Vienna, Akad. der Wiss., *Almanach;* Brussels, Acad. roy. des sci., *Regl.;* London, Royal Soc., *Yearbook*, and MacLeod, *Notes and Records, 26* (1971), 94; Gauja, *Fondations de l'Académie des sciences* (1917); Berlin, Akad.

ernment than on private donors, who made them very large gifts around the turn of the century. The increase was particularly important to French scientists because the terms of some of the new bequests allowed the Academy to spend money for work in progress or planned rather than as a prize for completed work.[47] Our table does not take into account informal support like the Berlin Academy's "loan" to Kayser of a Rowland grating bought for his use;[48] we suspect that if we could add such support and subtract prize money that was not spent on research, the German

---

between 1896 and 1905 the grant dispensed 750,000 marks, of which 19 percent went to medical, physiological, and pharmaceutical research, 3.3 to applied physics, and 5.8 to basic physical research. For French prizes granted in 1900 the comparable figures were: medicine, etc., 32 percent; engineering, 11 percent; basic physical research, 8.2 percent. *CR, 131* (1900), 1041 ff.

[47] For example, some 800,000 marks from Debrousse and 80,000 from Henri Becquerel. *CR, 131* (1900), 1040; *147* (1908), 483. For donors see Darboux, *Éloges* (1912). See supra, note 38 to section B.

[48] Kayser, "Erinnerungen" (1936), p. 146. According to Voigt, the Berlin Academy also loaned an electromagnet to Runge. Göttingen, U., *Phys. Inst.* (1906), p. 41. Cf., however, note 80, infra.

Notes to Table C.4 *(continued)*.

der Wiss., *Abhandl.;* Göttingen, Ges. der Wiss., *Nachr.;* Tokyo, Imperial Acad., *Proc., 1* (1912). Italy, Min. dell'ed. naz., Direzione generale delle accademie, *Accademie e istituti di cultura. Fondazioni e premi* (1940), gives a list of all academic prizes and winners since the mid-nineteenth century. Academies rewarding physicists (numbers indicate value of prize in 1000's of lire): Milan, Ist. lombardo (p. Cagnola, 1-2.5); Rome, Acc. dei lincei (p. reale, 10; p. ministeriale, 4; and, after 1911, an annual p. Alfonso Sella, 1); Rome, Società italiana delle scienze (p. Matteucci, 0.2 [a gold medal], not included in our counts); Turin, Acc. delle scienze (p. Bressa, c. 10; p. Vallauri, c. 30).

[b] Calculated for France over 1890-91 to 1912-13; for Germany and Austria over 1891-1900 to 1901-10; for the U.K. over 1895-99 to 1900-05.

[c] "Ray" physics includes cathode rays, X rays, radioactivity, and magneto-optics. Cf. note 45.

[d] Schück, et al., *Nobel* (1962), pp. 644, 666. We give the prize's 1901 value, although this increased; cf. Eve, *Rutherford* (1939), p. 184. We count physics prizes, excluding, e.g., Rutherford's chemistry prize. Since the awards began after 1900, the Nobel prize receipts have *not* been carried forward to Tables C.1 and I.

[e] Of this sum 3,200 marks were given for theoretical physics prizes. The Academy gave another 20,500 in these ten years as prizes to foreigners for "ray" physics. Gauja, *Fondations de l'Académie des Sciences* (1917).

[f] Very rough estimates concocted from Hull, *Funds Available* (1921); Baker in Carnegie Inst., *Report Executive Comm.* (1902), pp. 248-50; True, *History N.A.S.* (1913); Pickering, *Science, 13* (1901), 201-02; Brown, "Need of Endowment" (1890), p. 1065; Miller in *Sci. and Society in the U.S.* (1966), pp. 191-221. We have not included the grants by the Carnegie Institution of Washington in as much as these began only after 1902. See notes 73 and 75.

[g] The first figure refers to the prizes counted in column 1; the second to the premio Matteucci, 1896-1905 (see note a, above).

academies' funding of physics research would not fall below the French. The British remained far in the rear.

France had a special institution, the École pratique des hautes études, designed to provide funds, as determined by committees of scientists, to laboratories where advanced students could mingle with professors in small cooperative groups dedicated to pure research.[49] Had it functioned as intended, the École would have put some 60,000 marks a year at the disposal of researchers in physics and chemistry. Instead, bureaucrats captured the École and used it to dispense funds at fixed annual rates to teaching laboratories in existing institutions.[50] It did supply scholarships and paid for Lippmann's laboratory for pre-doctoral students, but since it operated much like other funding agencies of the Ministry of Public Instruction, its contributions have been incorporated in Table C.2. In 1902 the French again tried to create a source of direct funding for research, the Caisse des Recherches Scientifiques. Prominent physicists sat on the committees which dispensed over 110,000 marks a year. But the aims of the parliamentary proponents of the Caisse guaranteed that almost all its money went to biological and medical research and to public health measures; essentially none of it went to physics.[51]

In the United States, the National Academy of Sciences, the American Academy of Arts and Sciences, and the Smithsonian Institution all had some money for research. But they preferred other sciences, and particularly astronomy, over physics, which received relatively little.[52]

Another very considerable source of support for universities, particularly in Europe, was local government or, in Italy, a *consorzio* of local governments and business. In Germany in the decade 1891–1900 nine of the twenty-seven towns possessing higher schools made money contributions, in addition to grants of land; the sums ranged, for special grants, from 23,000 marks (Greifswald) to 1,300,000 (Darmstadt), and, for annual allowances, from 5,000 (Karlsruhe) to 80,000 (Darmstadt).[53]

[49] Liard, *Enseignement supérieur, 2* (1894), 286–95; Duruy, *Notes et souvenirs, 2* (1902), 300.

[50] Wallon in *RIE, 25* (1893), 474–76; cf. *RIE, 26* (1893), 258–59; Lexis, *Hochschul-Nachrichten, 11* (1901), 221.

[51] France, *Journal officiel* (1901), p. 1443; France, Min. instr. publ., Caisse Rech. Sci., *Rapports* (1905); Foville, *Revue Sci., 49* (1911), 385–88.

[52] See notes to Table C.4. We may here note the Rumford Fund of the American Academy of Arts and Sciences, a steady supporter of radiation research (c. 25 grants for physics, average value 1,200 marks, in the decade 1896–1905). Amer. Acad. Arts Sci., *Rumford Fund* (1905), pp. 14–21.

[53] *Hochschul-Nachrichten, 11* (1901), 198. The total for special grants was 2,753,000 marks; for annual, 160,000.

For France we have indications only: there were substantial contributions from the city of Marseille for a new science building, and 180,000 marks from Nancy, 1895-1905, for its university's physics institute.[54] In England in 1909-10 the percentage of total university income deriving from *annual* grants from local authorities ranged from 1 to 37 percent, with an average of 15, as was the case with endowments.[55] In Italy the *consorzio* might mean the difference between death and stagnation for a university of the second class, and between stagnation and growth for one of the first.[56] We presume that most of the portion of this money that reached physics has found its way into our tables.

c. Direct Philanthropy

Unlike their continental equivalents, private donors in Britain tended to give their money directly to the recipient or to his university,[57] rather than channeling it through a central body. Much of this money went either into endowed chairs, equipment funds, and general university funds, or into building and equipping new laboratories; consequently it is incorporated in Table C.2 or Table D.1. Table C.1 estimates the remainder; the values may well be too low, for no doubt many small gifts have eluded us. We have entered a little over double the remainder.

These small grants were important. Unlike endowed chairs and new buildings, they yielded income which could be used directly for research. Grantors came from many directions. The Goldsmith's Company of London gave 20,000 marks to aid Dewar's low temperature researches;[58] the principal auditors at Kelvin's lectures in Baltimore got up a fund to buy him a Rowland diffraction grating (which would have cost about 1,000 marks);[59] a shipowner and personal friend of Lodge supported the

54 France, Min. instr. publ., *Enquêtes,* Nr. 101 (1911), p. 70; Nancy, U., *Cinquantenaire* (1905), pp. 135-36.

55 G.B., Board of Ed., *Reports* (1909-10), pp. xx-xxi.

56 Ferraris, *Inst. veneto di scienze, Atti, 61,* pt. 2 (1901-02), 305-14, and *Cinque anni* (1922), pp. 40, 47-48; Genoa, U., *Univ. di Genova* (1923), pp. 83-96. Some sums supplied by *consorzi:* Padua, 26,000 marks for physics, zoology, surgery, and chemistry in 1906-07 (*Annuario,* p. x); Pavia, 1,300 marks for physics in 1899-1900, 2,100 in 1909-10 (*Annuario*); Turin, 2,600 in 1898, 1,800 in 1903-04 (*Annuario*). For details of the *consorzi* contributing to the several universities, see: Italy, Min. pubbl. istr., *Monografie delle università* (1911-13), passim.

57 We may mention the physical institutes at Manchester, Trinity College Dublin, and Oxford (Electrical Laboratory), financed by subscription, by an individual donor (Viscount Iveagh), and by a single corporation (The Drapers' Company of London), respectively. Sources in notes to Table D.1.

58 *Nature, 53* (1896), 425-27.

59 Thompson, *Kelvin, 2* (1910), 838. In 1895 the University of California paid

ether drag experiment which the Royal Society also aided;[60] shortly after the turn of the century, when liquid air plants suddenly became "indispensable" pieces of expensive apparatus, private donors quickly provided them at the major laboratories.[61] The universities of Scotland all benefited from the enlightened philanthropy of Neil Arnott and his wife, whose endowments for research and prizes in physics, given between 1869 and 1874, were worth about 165,000 marks in 1900.[62]

In Germany physicists enjoyed private benefactions on about the same level as in Britain. Often the gifts were in kind, like the 20,000 marks worth of liquid air that Siemens furnished Nernst,[63] and the expensive spectrograph mountings presented the Bonn and Göttingen institutes by Krupp.[64] But more commonly donations came via an intermediary—if not an academy, then a special fund associated with a university. One thinks first of the Carl Zeiss-Stiftung, established by the physicist Abbe at the University of Jena, and of the Göttinger Vereinigung für angewandte Physik und Mathematik.[65] But there were many other intermediaries, and significant private donors as well.[66] The Germans also had a few private or

---

840 marks for a 5″ Rowland grating: "This was (and continued to be) Rowland's standard price." Birge, "Physics Dept. Berkeley," *1* (1966), V, 30.

[60] Lodge, *Past Years* (1931), 195-99.

[61] Liquid air plants were donated to the physics institutes at Nancy, Manchester, McGill, Cambridge, Liverpool, Adelaide, Göttingen, Bonn, Chicago, infra, Section C.3. Note that much of this philanthropy was in kind; even Cambridge had trouble attracting endowment money around 1900. Cf. J.W. Clark, *Endowments of U. Cambridge* (1904), pp. 638-46; Rothblatt, *Revolution* (1968), pp. 255-57.

[62] Compiled from the universities' reports in G. B., Parl., H.C., *Sess. Papers*, 1902, *81*, 235-400, and Royal Soc. Lond., *Proc., 25* (1877), xiv-xviii. Edinburgh received another endowment for research in 1907, some 47,000 marks bringing about 3 percent. Edinb., U., "Stat. Report," in G.B., Parl., H.C., *Sess. Papers*, 1911, *72*, 89; cf. Edinb., U., *History* (1933), pp. 422-26, which reports a higher figure (82,000 marks).

[63] Birkenhead, *Lindemann* (1961), p. 36.

[64] Konen in *Zs. für wiss. Photogr., 1* (1903), 325-42; Riecke in Göttingen, U., *Phys. Inst.* (1906), p. 65; Voigt in *ibid.*, pp. 41-42.

[65] Wuttig, "Zeiss-Stiftung" (1930); Schomerus, *Zeiss-Stiftung* (1955); Auerbach, *Abbe* (1918). For the Göttinger Vereinigung see Manegold, *Univ., T.H. und Industrie* (1970), and Göttingen, U., *Phys. Inst.* (1906), pp. 19, 36, 198, et passim. For Zeiss's contributions to physics at Jena see Lexis, *Univ., 1* (1904), 585; for the Vereinigung's to Göttingen, e.g., 2,500 marks to study heat radiation and 10,000 for applied electricity given in 1908-09, see Göttingen, U., *Chronik*.

[66] E.g., the Akademische Gesellschaft at Freiburg, with assets of around 100,000 marks in 1907. Baumgarten, *Univ. Freiburg* (1907), pp. 146-47. See also Frankfurt, Phys. Verein, *Neubau* (1908); and (re the importance of Stiftungen in outfitting the Bonn physics institute) Konen, *Ges. U. Bonn* (1933), p. 353. Many gifts and legacies in Germany will be found listed in *Hochschul-Nachrichten*, e.g., *11* (1901), 100, 182,

industrial contributors to the construction and furnishing of new laboratories.[67]

In France the total of gifts to physics given privately or through the university was small in 1900, but growing rapidly. We may mention the good luck of a professor at the University of Nancy who, after lamenting the want of a liquid air machine during a public lecture, was immediately given 4,000 marks to buy the apparatus by an auditor touched by his plight.[68] The largest private donation for science in the decade after 1900 was probably the 3 million marks left the Paris faculty of sciences by G.A. Commercy in 1908, from which basic physical research derived some 42,000 marks in the next two years alone.[69] France also had a few private institutional donors, particularly the Société d'Encouragement pour l'Industrie Nationale, which supported M. Curie's work on steel, and the Société des Amis de l'Université de Paris, which periodically distributed amounts of the order of 1,000 marks to assist physical research at the Sorbonne.[70]

In the United States private gifts to physics were most important, albeit difficult to compute. We may point to the 16,000 marks the Harvard physics department drew from an endowment fund in 1907, a year in which it also picked up 22,000 in gifts.[71] Private funding increased at

---

233. Some gifts to university physics institutes: Raky to the U. Bonn for a machine house (*Chronik* [1906], pp. 39-40); H. Hauswaldt to the U. Göttingen for magnets and spectroscopes, and von Boettinger 2,200 marks for spectroscopic equipment (*Chronik* [1906-07], p. 47); Hauswaldt again to the U. Breslau c. 7,000 marks for a Fabry-Pérot interferometer (*Chronik* [1907-08], p. 47); Kurator to the U. Bonn for liquid air machine (Kayser, "Erinnerungen," p. 256); "private sources" to Nernst's institute, U. Berlin, 25,000 marks in 1909 (Lenz, *Gesch. U. Berlin*, pp. 306-10).

[67]Göttingen, U., *Chronik* (1905-06), p. 44 (Boettinger and Krupp); Konen in *Gesch. U. Bonn, 2* (1933), 345-55 (Kurator U. Bonn); Frankfurt, Phys. Verein, *Neubau* (1908), p. 87 (Frankfurt electrotechnical industry, some 30,000 marks).

[68]Nancy, U., *Séance* (1900), pp. 30-31. Other French gifts and legacies through 1905 (very few specifically for physics) are listed in Picavet, *RIE, 50* (1905), 22-48; for private benefactions behind the Paris radium institute see Liard, "Les bienfaiteurs" (1913), pp. 341-42.

[69]Paris, U., Conseil acad., *Rapports* (1909-10), pp. 126-31; (1910-11), pp. 167-90; (1911-12), pp. 218-19. France, Min. instr. publ., *Enquêtes*, Nr. 101 (1911), pp. 46-47. In 1912 the fund bought Marie Curie a liquid air machine; *ibid.* (1912-13), pp. 200-02.

[70]Soc. d'Encour., *Bulletin, 97* (1898), 36; France, Min. instr. publ., *Enquêtes* Nr. 108 (1919), pp. 22, 97. The Amis gave, e.g., 800 marks to Lippmann's and 1,000 to M. Curie's laboratory in 1909-10 (*ibid.*, Nr. 101 [1911], pp. 46-47), and 1,600 to the former's in 1906 (Paris, Conseil acad., *Rapports* [1906-07], p. liii).

[71]Harvard U., *Treasurer's Report*, as given by Goldberg, "Physics at Harvard" (1962), p. 25; cf. McClenahan, *Science, 32* (1910), 295, announcing an endowment

least as rapidly in the United States as in other countries, for after 1900 the great philanthropic foundations began to give very substantial sums to science.[72] The Carnegie Institution of Washington, which started its grants in 1906, had given 32,000 marks towards basic physics research in the United States by the end of 1910.[73] The large and well-advertised scientific philanthropy in the United States impressed Europeans and suggested a new route to the pockets of their countrymen. "It would be very much in the interest of Germany and the supremacy of her science if . . . private persons would support scientific institutes by furnishing money or apparatus, as has long been the custom in other countries, and particularly in the United States." "The modern apparatus [that American universities receive from private sources] made my mouth water."[74]

After the turn of the century physics benefited more and more from philanthropy on an international scale. Carnegie gave large sums not only in America, but also in his native Scotland and in France, and not only to physics but to universities in general.[75] The Nobel Prize, over 150,000

---

of 840,000 marks, the income to be used for apparatus, supplies, and research assistants. At Columbia the Phoenix Trust provided 12,600 marks per year for physics research from 1903 and promised a raise to not less than 40,000 when other beneficiaries died. Hallock, *Columbia U. Quart., 5* (1903), 294.

[72]Curti and Nash, *Philanthropy in Higher Ed.* (1965); Goodspeed, *Hist. of U. Chicago* (1916), pp. 179-93.

[73]Compiled from Carnegie Institution, *Yearbook*. We do not include the much larger grants to solar physics. The American physicists were, however, quite disappointed by the Institution's policy of reserving most of its funds for the large projects in its own laboratories. Edward B. Rosa to A.G. Webster, 12 March 1906 (Niels Bohr Library, American Institute of Physics, N.Y.)

[74]Quotes from, respectively, Kayser in Bonn, U., *Chronik* (1902), pp. 58-59, and Ayrton in *Mosely Commission Reports* (1904), p. 29. In general the two decades before the war were distinguished as a period of great European interest in, and, perhaps for the first time, emulation of, American philanthropic methods and institutions for the advancement of science. This was preeminently so in Germany and is clearly reflected in the supporting institutions founded in this period, notably the Kaiser-Wilhelm-Gesellschaft zur Förderung der Wissenschaften and the Universität Frankfurt, as well as by the numerous study tours etc. E.g., Max-Planck-Ges., *50 Jahre* (1961), pp. 80-94; Böttger, *Amer. Hochschulwesen* (1906); Meyer, "Studies of Museums etc." (1905), pp. 321-22, and passim. German scientific philanthropy itself was the object of emulation of other countries: see, e.g., Kalinka, "Österreichische Forschungsinstitute," p. 52; G.B., Board of Ed., *Reports* (1909-10), pp. iii-vi, which also expresses envy over French endowments. Cf. note 4 to Section A, supra.

[75]*Nature, 80* (1909), 20-21; Martin, *RIE, 51* (1906), 18-20. In 1907 Carnegie gave the Sorbonne an endowment of 200,000 marks, yielding 10,000 a year, to provide

marks, had an impact on physics almost from the start; the Curies used part of theirs to hire an assistant in the laboratory provided them by the government after the prize had made them famous; Rayleigh gave part of his to help finance a large new wing at the Cavendish.[76] The Solvay Institute for Physics, founded in 1913, immediately became an important source of funds for up-to-date research, for which it gave some 17,000 marks in its first year of operation.[77]

3. Aids to Research

The ordinary and extraordinary funds we have reported procured the academic physicist two kinds of aids and instruments of research: extramural purchases and, what was perhaps more important, intramural technical and menial assistance.

In 1900 most physical laboratories had at least one mechanic or machinist and one menial servant (Table A.4). Few such men existed in 1870; but thirty years later most larger institutes had three to six, and, as Table C.5 suggests, academic physics as a whole then employed nearly three hundred. Most of these men, who served only to help set up apparatus for lecturer and students, sweep, and shovel coal, merited the title *garçon de laboratoire* or *Diener*. But many were true mechanics, expert instrument makers able to blow glass and do precision machine work. Often they built not only apparatus for the undergraduates but also research instruments, especially electrometers and astatic galvanometers, which were too delicate to ship.[78] In some places, for example Leipzig, the head

---

fellowships (Bourses Curie) for researchers attached to the Curie laboratory. Appell, *Revue de Paris, 17*, no. 6 (1910), 111; Paris, U., Conseil acad., *Rapports* (1906-07), pp. 98-99; Carnegie Endowment for Intl. Peace, *Benefactions of Carnegie* (1919), pp. 311-21. We may also mention Carnegie assistantships at U. Aberdeen (1,500 marks) and at U. Lund (1,100); annual grants for apparatus at U. Aberdeen (1,200); a one-time grant for a physics laboratory (128,000) and apparatus (12,500) at U. Edinburgh. Aberdeen, U., "Abstract of Accts," and "Stat. Reports," 1900-10; Edinb., U., "Stat. Report," in G. B., Parl., H. C., *Sess. Papers*, 1909, *69*, 612; Oslo, U., *Festskrift 1911-1961, 1* (1961), 510.

[76] E. Curie, *M. Curie* (1937), pp. 211, 236-39; Rayleigh, *Rayleigh* (1968), pp. 314-15. See Table C.4 for Nobel Prizes.

[77] Pelseneer, "Hist. Solvay" (1962), pp. 20-30. Much information about the early Solvay Institute appears in the Lorentz correspondence (AHQP); e.g., in the exchanges with Knudsen (M/f 12).

[78] H.F. Weber, "Bericht . . . 1910." Marburg, U., *Chronik* (1909), p. 44, and (1910), p. 47, also lists the principal apparatus constructed in the workshop. The in-

TABLE C.5

Institute Mechanics and Servants circa 1900[a]

	Mechan-ics	Ser-vants	Total, 1900	Change 1900–10	Percent per annum
Austria-Hungary	9	17	26	2	1
British Empire			53	20	4
France	10	20	30	3	1
Germany	27	31	58	20	4
T.H.	11	7	18	5	3
Univ.	16	24	40	15	4
Italy	20	21	41	9	2
Netherlands			13	4	3
Scandinavia			7	3	4
Switzerland	3	7	10	5	5
United States	(15)	(25)	(40)	(30)	(8)
Total			280	100	4.5

[a]Based upon the data presented in Table A.4. The figures, apart from those for Italy, Switzerland, and the German universities, are essentially guesses; those for the United States merely guesses.

stitute mechanics were doubtless responsible for a considerable part of the "marvelous Sammelung [sic]" of the physical institute of the University of Berlin (Crew, "Diary, 1895," entry 13–15 July), which by 1909 numbered some 1,800 items. Rubens in Lenz, *Gesch. U. Berlin, 3* (1910), 296. Cf. Vienna, Akad. Wiss., *Almanach* (1891), p. 178; Rayleigh, *Rayleigh* (1968), pp. 103–04; Küchler, "Phys. Labs. in Germany" (1906), p. 205. At tiny Clark University "one of the most important adjuncts of the department [is] the workshop, in which a skilled mechanic is constantly employed in the constructing of apparatus for research." A.G. Webster in *Clark U. 1889–1899*, p. 96, and in *Report of the President* (1902), p. 35. Likewise Harvard U., *Guide Book* (1898), p. 86; Pennsylvania, U., *Guide* (1904), pp. 16–17. Konen, *Reisebilder* (1912), p. 50, found this to be characteristic of the U.S. institutes–which were correspondingly well equipped with machine tools–due to the high cost of extramural work. In 1900, however, the Columbia University physics department, despite its staff of twelve academics, had no mechanic: "at present all the apparatus used in research work is constructed by the professors and assistants themselves with such skill as they may individually possess." *Columbia U. Quarterly, 3* (1900), 196.

mechanic supervised a small establishment of assistants, complete with apprentices.[79] The mechanic's principal responsibility was the construction and upkeep of apparatus for the teaching laboratory and demonstration. Whatever additional time he had was at the disposition of the director alone; junior faculty constantly complained about the amount of technical assistance available to them.[80]

Although in this period the physicist continued to improvise apparatus for himself out of "string and sealing wax," "scraps of cardboard or of copper wire," or the tobacco cans favored by Rutherford's school, and continued to pride himself upon doing so,[81] increasingly he had recourse to the growing number of commercial manufacturers offering an ever wider range of ever more expensive instruments.[82] In the past such purchases had been almost exclusively for apparatus which, although em-

[79]Wiener in Leipzig, U., *Festschrift, 4*, Pt. 2 (1909), 59. The physical institute at T.H. Darmstadt had two apprentices in 1900, three in 1910. *Programm* (1900-01), pp. 7-13, and (1909-10), pp. 12-18. Such apprentices are not included in Table A.4. Often the institute mechanic was especially skilled or experienced in constructing a particular type of apparatus, perhaps one invented by his director, which might be sold or traded for the specialty of another laboratory. E.g., Paschen to Johann Koenigsberger, 16 March 1913 (Preussiche Staatsbibliotek, Berlin). Willy-nilly the physical institute served to train mechanics for the burgeoning instrument industry; at Leiden Kamerlingh Onnes capitalized upon this circumstance by establishing a formal apprenticeship program. Leiden, Natuurkundig Lab., *Gedenkboek* (1904), pp. 60-68; Leiden, Natuurkundig Lab., *Gedenkboek* (1922), pp. 78-86.

[80]Note the large amount of work done for J.J. Thomson by his assistant, E. Everett (Cavendish Lab., *History*, pp. 82, 197); the scientific standing of Henri Becquerel's factotum, Matout, who published papers and abstracts in his own name; and the policy of the mechanic at Manchester, who placed the repair of Rutherford's motor car above making apparatus for the research men (Heilbron, *Moseley*, p. 181). Also Kay, *Nat. Phil., 1* (1963).

[81]Broca, *RGS, 19* (1908), 802-13; Hahn, *Autobiog.* (1966), p. 32; *Nature, 65* (1902), 587-90.

[82]For the availability of, and dependence on, commercial cathode ray tubes, etc., see Glasser, *Dr. W.C. Röntgen* (1958), pp. 22, 87-88; Glasser, *Röntgen* (1959), pp. 279-86; Röntgen, *Briefe an Zehnder* (1935), pp. 28-29, 52-54; Deslandres, *CR*, *124* (1897), 678-81; Moseley to Rutherford, 7 Dec. 1913 and 5 Jan. 1914, in Heilbron, *Moseley*. Much information on equipment may be found in *Le Radium* and *Zeitschrift für Instrumentenkunde*, and in manufacturers' catalogs, e.g.: Société genevoise, *Prix courant* (1896); Vereinigte Fabriken, *Liste Nr. 60* (1906); M. Kohl, *Price List Nr. 50* (1911); Leiss, *Instrumente der Firma R. Fuess* (1899); Société des Lunetiers, *Instruments* (1908). A list of instrument manufacturers appears in the *Adressbuch* (1909), pp. 153-99, and statistics of the value of the product of the (preeminent) German fraction are given in Paris, Exposition universelle, 1900, *Catalog of the Exhibition of German Mechanicians*, pp. 8-9. Cf. Rees, *Science, 12* (1900), 777-85, and De Metts, *Inzhener, 23* (1899), 462-66, 498-506.

ployable in research, was intended primarily for instruction in lecture hall or laboratory.[83] But from the 1890's on an increasing proportion of the apparatus bought was intended largely or exclusively for research.[84]

Unquestionably the most novel and expensive addition in this period to the instrumental facilities of the well equipped physical institute was an apparatus for the production of liquid air. The entire plant, comprising liquefier, compressor, and motor, cost between 4,000 and 6,000 marks, of which 1,000–1,500 marks went for the liquefier per se. Several systems were patented and marketed from 1895 onward; typical plants yielded about one liter per hour, but some high powered American apparatus exceeded that several-fold.[85] Our earliest instance of the purchase of a commercially manufactured apparatus is G. W. A. Kahlbaum's present to the physical institute of his own university, Basel, in 1897.[86] Such installations soon came to be regarded as a necessity for every modern physical laboratory.[87] Within a decade of Kahlbaum's precocious gift Nancy, Montpellier, Giessen, Hamburg, Breslau, Danzig, Leipzig, Kiel, Freiburg, Stuttgart, Tübingen, Manchester, McGill, Liverpool, the Cavendish, Adelaide, Cornell, Minnesota, Chicago, Pennsylvania, and presumably many other physical institutes acquired them;[87a] in the seven years remaining before the war

[83] Lists of all apparatus purchased for the T.H. Karlsruhe physical institute in the forty years prior to 1911 are given, with prices, by Lehmann, *Phys. Inst. Karlsruhe*, pp. 70–78. Lists of current purchases, but without prices, appear in Halle, U., *Chronik* (1901–06); Greifswald, U., *Chronik* (1900–01); Breslau, U., *Chronik* (1896–1903); Uppsala, U., *Redogörlse* (1890–1910); Lund, U., *Årsberättelse* (1890–1910).

[84] For descriptions and/or lists of apparatus see Hallock, *Columbia U. Quarterly, 5* (1903), 295; Fowler, Phys. Soc. Lond., *Proc., 24* (1912), 168–71; Manchester, U., *Phys. Lab.* (1906), pp. 25–38; Chicago, U., *President's Report* (1902–04), 215; McGill, U., *Opening* (1893), p. 67. The equipment list of the McGill laboratory included four different electrometers, ostensibly for instruction; they were used in the fundamental researches of Rutherford and Soddy. Romer, *Discovery of Radioactivity* (1964), pp. 38, 95, 131. Cf. Henry Crew's impressions of the ETH physical institute ("Diary, 1895") quoted by McCormmach in "Editor's Introduction" to volume Four of *Hist. Stud. Phys. Sci.*.

[85] Allen and Ambler, *Phys. Rev., 15* (1902), 181–87; Hudson, *Liquid Air Plant* (1908).

[86] Thommen, *U. Basel* (1914), p. 117. Our next instance was a gift to the physical-chemistry laboratory at the Sorbonne. Picavet, *RIE, 50* (1905), 22–40.

[87] Küchler, "Phys. Labs. in Germany," pp. 203–04.

[87a] Indeed in that decade Linde supplied liquefiers also to the physical institutes of U. Moscow, U. St. Petersburg, U. Charkov, U. Kasan; T. H. Budapest, U. Krakau, U. Vienna; U. Munich, U. Berlin, T. H. Darmstadt, T. H. Berlin, U. Königsberg, T. H. Aachen, T. H. Hannover; Copenhagen Polyt.; E. T. H. Zürich, U. Fribourg; U. Nancy, U. Lille. "Verzeichnis der bisher gelieferten . . . Luftverflüssigungs–Anlagen System Linde" (1927), from Archive, Linde AG, Höllriegelskreuth. [Note added in proof.]

the Bonn, Marburg, Vienna, Copenhagen, Edinburgh, Glasgow, Oxford, Urbana, Princeton, Evanston, M.I.T., and Curie laboratories were similarly outfitted.[88]

These installations were indispensable for a wide range of important and up-to-date investigations besides those intrinsically tied to low temperature phenomena.[89] Liquid air found its widest employment in fundamental physics research as part of a revolution in vacuum technology.[90] Dewar's technique of attaining high vacua by adsorption of gases onto cooled charcoal required not only liquid air, but also high-speed pumps capable

[88]Nancy, U., *Séance de rentrée*, 1899 (1900), pp. 30-31, but it is not certain that the *physical* institute was the recipient; Montpellier, U., *Livret de l'étudiant* (1902-03), p. 122; W. Wien, *Phys. Zs. 1* (1900), 157, for Giessen; Voller in *Hamburg in naturwiss. Beziehung* (1901), p. 211; Breslau, U., *Chronik* (1900-01), p. 33; Danzig, T.H., *Festschrift zur Eröffnung* (1904), p. 21; Wiener in Leipzig, U., *Festschr., 4*, pt. 2 (1909), 46; Küchler, "Phys. Labs. in Germany," pp. 203-04, for Freiburg i. Br.; *ZBBV, 23* (1903), 158, for Kiel; Koch, *Phys. Zs., 12* (1911), 830, for Stuttgart; Württemberg, Kammer der Abgeord., *Verhandl.,* 36. Landtag (1904-05), Beilagenband 2, Heft VI, 240-41, for Stuttgart and Tübingen; Manchester, U., *Phys. Lab.* (1906), p. 36; McGill U., *Annual Report* (1902-03), p. 35, and Eve, *Rutherford* (1939), p. 92; *Nature, 71* (1904), 63-65; Cavendish Lab., *History* (1910), p. 191; W.H. Bragg to Rutherford, Adelaide, 2 August 1907 (Rutherford Papers, Cambridge Univ. Library); Allen and Ambler, *Phys. Rev., 15* (1902), 181-87; Erikson, "Univ. of Minnesota Dept. of Physics," p. 20; Chicago, U., *President's Report* (1902-04), 215-17; Pennsylvania, U., *Guide* (1904), p. 17.

1907-14: Copenhagen, *Aarbog* (1909-10), p. 1411; Bonn, U., *Chronik* (1910), p. 49, and Kayser, "Erinnerungen" (1936), p. 256; Schulze in Marburg, U., *1527-1927*, p. 759; Austria, Min. f. Kultus u. Unterr., *Neubauten* (1913), p. 6, for U. Vienna; Glasgow, U., "Stat. Report," in G. B., Parl., H. C., *Sess. Papers*, 1910, *72*, 15; Edinburgh, U., *ibid.*, p. 732; Moseley to Rutherford, Oxford, 7 Dec. 1913, in Heilbron, *Moseley*, p. 216; Hudson, *Liquid Air Plant* (1908); H. Crew (Evanston) to W.F. Magie (Princeton), 30 April 1910 (Crew Papers, AIP); McClenahan, *Science, 32* (1910), 293; M.I.T., *President's Report* (1912), p. 106; Paris, U., Conseil acad., *Rapports* (1912-13), pp. 200-02. The absence of many of the largest and most active laboratories from this list is not necessarily due to its incompleteness; as late as 1911 Harvard, for example, was purchasing liquid air from a commercial supplier in Buffalo, New York. Goldberg, "Physics at Harvard" (1962), p. 7. Göttingen—and not its physical institute but its institute for applied mathematics and mechanics—obtained a liquefier only in 1913. *Chronik* (1913), p. 52. In 1905 liquid air was advertised for sale in Berlin in 2 liter flasks at 2 marks per liter. Kausch, *Flüssige Luft* (1905), advertisements at end.

[89]Cavendish Lab., *History* (1910), p. 101; Thomson, *Recollections* (1937), pp. 354, 374. Rutherford, *Collected Papers, 1,* 528. The liquid air apparatus was by no means solely an instrument of research. The University of Minnesota plant provided the substance for "liquid air lectures" which the professor of physics delivered around the state. Erikson, "U. of Minn. Dept. of Physics," pp. 20, 259.

[90]For vacuum technique see Dushman, *Vacuum Technique* (1949), pp. 137-90; H. Gaede, *Wolfgang Gaede* (1954).

of maintaining an intrinsically leaky apparatus at about 10^{-2} mm Hg. Here entered a second line of advance, the development of mechanical vacuum pumps driven by electric motors. In the early years of the century the best of these, the Geryk pump, which could reach 2×10^{-4} mm Hg, cost about 1,000 marks;[91] further improvements, particularly by Wolfgang Gaede, brought down the vacuum, but not the cost, by an order of magnitude.[92]

Another standard piece of laboratory apparatus which was rapidly increasing in power and expense in this period was the electromagnet. In 1900 Carl Runge and Friedrich Paschen paid 1,400 marks for a half-ring magnet of the du Bois type, the latest design, with which they reached fields of 25,000 gauss in their studies of the Zeeman effect.[93] The magnet constructed by Pierre Weiss in 1907, capable of over 40,000 gauss, soon made all previous types obsolescent. Low powered imitations appeared in the instrument catalogs at 2,000 marks; when Paschen had one built for himself in 1913 he probably paid twice that.[94] Other important and expensive research apparatus were: high voltage storage batteries of 1,500 elements, 4,000–5,000 marks;[95] large induction coils, upwards of 400 marks; Wehnelt or rotating mercury interrupters for the same, 100–200

[91] Vereinigte Fabriken, *Liste Nr. 60* (1906), p. 295. Lehmann, *Phys. Inst. Karlsruhe* (1911), pp. 75–78, lists a Geryk pump purchased in 1903 for 354 marks, presumably hand operated: Hahn-Machenheimer, *Zs. f. phys. u. chem. Unterr., 14* (1901), 285–87.

[92] M. Kohl, *Price List Nr. 50* (1911), Nr. 52992. Gaede's rotatory mercury pump, introduced in 1905, cost about 400 marks; but since it required a fore pump and 20 kg of mercury, a full installation would have come close to 1,000 marks. Cf. Dushman, *Vacuum Technique* (1949); [Gaede], *Zs. f. Instrumentenkunde, 27* (1907), 163–65; Karlsruhe, T.H., *Festschr.* (1950), p. 48.

[93] Du Bois, *Zs. f. Instrumentenkunde, 19* (1899), 357–64; Akad. der Wiss., Berlin, *Sb.* (25 Oct. 1900), p. 928; Runge and Paschen, Akad. der Wiss., Berlin, *Abhl.* (1902), p. 5. These magnets were made by Hartmann and Braun, who also supplied a similar one to the physical institute of Breslau in 1900 (*Chronik* [1900–01], p. 33) and, for 2,000 marks, to that at Halle in 1901 (*Chronik* [1901–02], p. 43). Badash, *Centaurus, 11* (1966), 236–40, finds that c. 1900 the best equipment generally available gave fields of 8–10,000 gauss.

[94] Weiss, Soc. fran. de phys., *Bull., 35* (1907), 124–40; M. Kohl, *Price List Nr. 50* (1911), Nr. 62309. In December 1912 the Württemberg Kultusministerium approved Paschen's request for a special grant of 3,000 marks for this purpose (U. Archiv, Tübingen, 117/904); it seems unlikely that this was the entire cost. Kayser also purchased one for Bonn. *Chronik* (1910), p. 49. The culmination of this evolution was to have been Aimé Cotton's mammoth to which the University of Paris contributed 40,000 marks in 1913–14. France, Min. instr. publ., *Enquêtes*, Nr. 108, p. 96; E. Cotton, *A. Cotton* (1967).

[95] F. Paschen to L. Graetz, 22 July 1901 (Deutsches Museum).

marks; X ray tubes, 10–40 marks;[96] a 5-inch Rowland ruled diffraction grating, 800–1,000 marks;[97] a precision optical spectrometer reading to 1″, 1,500 marks;[98] a precision vacuum spectrograph for the ultraviolet, 3,000 marks; a large (4 cm X 4 cm) fluorite prism for the same, 2,000 marks;[99] a measuring machine good to 1/100 mm for evaluating spectrograms, 300–600 marks.[100]

Finally, a word about that most interesting and expensive research "material," radium. In the early years of the century the chief supplier of bona fide researchers was Friedrich Giesel, who produced radium bromide in the laboratories of the chemical firm which employed him, and sold it for a nominal price which rose from 10 marks per milligram in 1902 to 25 marks in 1903, 50 marks in 1905, and 100 marks per milligram circa 1906, when he discontinued the practice.[101] Thereafter the physicists were at the mercy of the Austrian state monopoly and the commercial suppliers, obliged to compete with the clamoring physicians. By 1908 they had to pay 250–300 marks per milligram, and by 1914, 350–400 marks.[102] At the same time the stock of the material required for competitive research continued to rise. In 1903 20 or 30 mg was a very large quantity. By 1906, when "no physical laboratory could be considered complete without the means for carrying on research in this branch of science," 60–70 mg was a good supply.[103] By 1914 there was probably no single piece of research apparatus in the larger physical institutes approaching the value of their radium stocks.

96 M. Kohl, *Price List Nr. 50* (1911), Nr. 62513-62702; Glasser, *Röntgen* (1959), pp. 88, 302.

97 Birge, "Physics Dept. Berkeley," *1* (1966), v (10).

98 Société genevoise, *Prix Courant* (1896), pp. 77–78.

99 Paschen, 1 Feb. 1912 (U. Archiv, Tübingen, 117/904).

100 Société genevoise, *Prix Courant* (1896), p. 14; Crew, "Diary 1895."

101 Röntgen, *Briefe an Zehnder* (1935), p. 95; Hahn, *Autobiog.* (1966), pp. 38–39; Hahn, *Selbstbiog.* (1962), pp. 29, 37; Eve, *Rutherford* (1939), p. 92; McGill Univ., *Annual Report* (1902–03), p. 35. Cf. note 34 to Section B, supra.

102 Hahn, *Autobiog.* (1966), p. 54; Hahn, *Selbstbiog.* (1962), p. 42; Eve, *Rutherford* (1939), p. 232; Tilden, *Ramsay* (1918), p. 169; Erikson, "Univ. of Minnesota Dept. of Physics," p. 37; S. Meyer to Rutherford, 17 Feb. 1914 (Rutherford Papers, Cambridge Univ. Library). In 1911 the Austrians valued the radium and other radioactive preparations of the Vienna Radium Institute at c. 1.7 million marks. Kalinka, "Österr. Forschungsinst.," p. 79.

103 Manchester, U., *Phys. Lab.* (1906), p. 34. Even Queen's University, Belfast had 60 milligrams, donated by Sir Otto Jaffe in 1904. *Pres. Report* (1908–09).

D. NEW PLANT

"Perhaps the most remarkable as well as promising fact relating to physical science at the close of the nineteenth century is the great and rapidly increasing number of well-organized and splendidly equipped laboratories in which original research is systematically planned and carried out."

T. C. Mendenhall, *New York Sun,* 17 Feb. 1901.

"Various circumstances, but above all the transformation of the pursuit of science into a large-scale industry [des wissenschaftlichen Betriebes in einen Grossbetrieb] in the last few decades . . . confers great current importance upon architectural questions."

Austria, Min. f. Kultus u. Unterr., *Neubauten* (1913), preface.

The above two quotations, which represent the viewpoints respectively of the academic physicist delighted with the new big physics and of the government planner concerned to fund it, may point up the great importance and novelty of the last items to be added to our survey of the physics enterprise circa 1900: the funds invested and the objects intended in the construction of physical laboratories during the quarter century from 1890 to 1914.

1. Rates and Causes of Laboratory Construction

In Table D.1 we list the physical laboratories affiliated with our set of higher schools (apart from Russia and Japan) which were constructed or enlarged in the years 1890–1914. The cost, documented or estimated, of each of these laboratories is given, as is a measure of the size of the facility thus placed at the disposal of our academic physicists. The documentation is fuller and our coverage better for those countries, notably those of Northern Europe, which tended toward structurally separate scientific institutes, poorer for France and the British Empire where in our period physics still commonly shared buildings with other sciences. In Table I we have sought to allow for our ignorance and, in general, for such expenditures on "new plant" as have escaped our attention, by augmenting appropriately for each nation the average annual expenditure yielded by the figures in Table D.1.[1]

In contrast with the costs, the floor areas given in Table D.1, even if we

[1] See Table I, note d.

knew them exactly for all the new facilities, are not a good measure of the total increase in the space available to academic physicists. Increase in space by accretion of or translation to rooms, wings, or entire buildings vacated by a neighboring discipline (usually chemistry) was common, especially outside Germany, but is quantitatively inestimable.[2] Against this unknown quantity must be set the unknown areas devolving upon other disciplines as a result of the removal of the physicists to the new quarters listed in Table D.1. Unquestionably there was a large net gain to physics, but we do not venture to guess how large. And of course the areas of Table D.1 do not include the space of the older laboratories still in service,[3] which cannot properly be ignored in a survey of the material dimensions of the enterprise.[4]

Table D.1 shows the British to have been the largest investors in bricks

[2] Instances of physics inheriting space vacated by chemistry: U. Paris, where in 1913 physics obtained the space previously occupied by chemistry in the new Sorbonne. France, Min. instr. publ., *Recueil des lois, 6* (1898-1909), 841-43; *idem., Enquêtes, 106* (1911-12), 85; Appell, *Revue de Paris, 17,* no. 6 (1910), 119. U. Liverpool in 1892. *Calendar* (1900-01), p. xxiii. U. Basel in 1910. Thommen, *1884-1913,* p. 118. Cornell U. in 1891. Howe and Grantham, "Phys. at Cornell," p. 6. New York U. in 1893. Jones, *NYU* (1933), p. 163. Clark U., where in 1903-04 physics moved into entirely new quarters occupying one wing of the former chemistry building. *Register* (1892), pp. 46-49; (1904), pp. 47-50.

[3] Surveys of the physical laboratories in one or more countries circa 1900: De-Metts, "Fizicheskie instituty" (1899), discusses Germany (pp. 462-66, 498-502), Switzerland (pp. 502-03), the Netherlands (pp. 503-06), and presumably other Western European countries in subsequent installments unavailable to us; Küchler, "Phys. Labs. in Germany" (1906), reports conditions observed in October and November 1903. Carl Junk, "Phys. Institute" (1905), describes the situation of the mid-1880's, when it was prepared for the first edition of the *Handbuch der Architektur.* (Those interested in chemical institutes are much more fortunate in E. Schmitt's article "Chemische Institute" (1905), pp. 236-382.) Likewise Robins, *Technical School and College Building* (1887). These older treatises consider fittings, wiring, ventilation, heating, etc., as well as floor plans and apparatus.

[4] Further descriptions of laboratories neither built nor rebuilt in 1890-1914: *British Empire:* Aberystwyth. Bd. of Ed., *Reports* (1894-95), p. 305. Bangor. *Calendar* (1900-01), pp. 220-21; Bd. of Ed., *Reports* (1899-1900), pp. 453-54. Birmingham. *Ibid.* (1894-95), p. 3. Bristol. *Ibid.* (1894-95), p. 34. Newcastle (Durham). *Ibid.* (1894-95), pp. 252, 268-70; (1906-07), p. 281. Sheffield. *History* (1955), p. 452. *France:* Lyon. *U. de Lyon, 1900,* pp. 11, 65. Montpellier. *Annuaire* (1902-03), pp. 122-24; *Fêtes du VIe cent.* (1891), pp. 43-45. Toulouse. *Annuaire* (1899-1900), pp. 92-94. *Italy:* Modena. *Annuario* (1899-1900), p. 178. Perugia. *Perugia e suoi istituti biologici* (1895), pp. 99-100. Rome. *U. di Roma* (1927), pp. 201-05. *Scandinavia:* Lund. Leide, *Fysiska institutionen* (1968), pp. 123-24; Örtengren, *Historiske notiser* (1950), p. 98. *Switzerland:* Bern. *Die naturw. Institute* (1896), pp. 5-9.

TABLE D.1

Physics Laboratories Constructed 1890–1914

Date[a]	Wholly new quarters	Cost[b] 1000 marks	Floor[c] area $10^2 m^2$	Factor[d]	Enlargements, etc.
	Austria-Hungary				
1898	Prague Böhm. U.	620			
1899		(50)	7	7	Brünn T.H.
1904	Innsbruck U.	(170)		(4)	
1909	Vienna T.H.*	(250)	15	3	
1910	Vienna Ra. Inst.*	385	16		
1913	Vienna U.	1750	90	1.2	
	Belgium				
(1912)	Brussels U. Libre				
	British Empire				
1892		50	3	2	Sheffield U.
1893	London U. Coll.	(110)	11	(5)	
1893	Montreal McGill U.*	980	36		
1896		83			Cambr. Cavendish
1896	London R. Inst.	[e]780			
1898	Aberdeen U.	(100)	8		
1900	St. Andrews U.	(250)			
1900	Manchester Owens*	730	27		
1900		(40)	3		Leeds U.
1901		20	2		Nottingham U.
1901		(150)	10		Sydney U.
1901		(100)			Adelaide U.
1902	Nat. Phys. Lab.*	[e]600			
1904	Liverpool U.*	490	36		
(1905)	Birmingham U.	(900)		(10)	
1906	Dublin Trinity Coll.	340			
1907	London R. Coll. Sci.	3000		2	
1907	Edinburgh U.*	(400)	20		
1907	Toronto U.	1700	44		
1908		(150)	12		Leeds U.
1908	Glasgow U.	820			
1908		150	10		Cambr. Cavendish*
(1908)	Cardiff U.	(50)		5	
(1908)		100			Bristol U.

(Notes and sources for table on pages 95–100.)

TABLE D.1 *(continued)*

Date[a]	Wholly new quarters	Cost[b] 1000 marks	Floor[c] area $10^2 m^2$	Factor[d]	Enlargements, etc.
1910	Oxford U. Electr. Lab.	460			
1911	Dundee U.	270			
1912		110		2	Manchester Owens
1913	Belfast Queen's Coll.	150		3	
1914	Dalhousie U.	(300)	(15)	(3)	
	France				
(1893)		150	9	4.2	Toulouse U.
1894	Poiters U.*	(150)	10		
1895	Lille U.*	360	12		
1896	Paris U.*	800	25	22	
1900	Paris U. PCN	(100)			
(1902)	Paris Éc. Phys. & Ch.	(300)	(25)	4	
1903	Paris. Lab. d'essais	[e]960			
1904	Toulouse U.	(300)	(25)		
1905	Paris U. Curie Lab.	28	(2)		
1909	Nancy U.	300	28	1.5	
1911		(50)			Clermont U.
1912	Gif. Radioact. Lab.*	[e](200)			
1913	Marseille U.	(150)		4	
1914	Paris U. Ra. Inst.	280		2	
	Germany				
1890	Freiburg i. Br. U.*	200	20	2	
1890		93			Marburg U.
1891	Greifswald U.*	180	20		
1891	Halle U.*	296			
1892		(50)			Aachen T.H.
1893	Munich U.*	430			
1893	PTR Phys. Div.	[e]826			
1894		(60)			Jena U.
1894	Erlangen U.*	212	18		
1895	Darmstadt T.H.*	250	18	2.5	
1896		(50)			Hannover T.H.
1897		80			Berlin T.H.*
1898	Hamburg Staatslab.*	[e]240	30		
1899	Giessen U.*	350	18	1.5	
1899	Münster U.*	127	15		

Table D.1 (*continued*)

Date[a]	Wholly new quarters	Cost[b] 1000 marks	Floor[c] area $10^2 m^2$	Factor[d]	Enlargements, etc.
1900		(50)			Freiburg i. Br. U.
1900	Breslau U.*	364	36		
1901	Kiel U.*	238	24		
1902		(40)			Halle U.
1902	Jena U.	(250)			
1904	Leipzig U.*	1284	67		
1904	Danzig T.H.*	400	15	7	
1905	Göttingen U.*	430	32		
1907		64			Munich T.H.
1908	Frankfurt Phys. Ver.*	[e]200	15	7	
1908		125	(6)		Tübingen U.*
1909	Stuttgart T.H.*	433	26		
1910		(100)	1		Leipzig U.*
1910	Rostock U.	240			
1911		81			Breslau U.
1913	Bonn U.*	437	53		
1913	Heidelberg U.	791	(40)		
1913		102			Münster U.
1914		61			Kiel U.
1915	Marburg U.	423	25		
	Italy				
1893		8			Padua U.
1894		(150)			Pisa U.
1896		(30)	(3)		Pavia U.
1898	Turin U.	(400)			
1907	Bologna U.*	430	46		
	Netherlands				
1892	Groningen U.*	(300)	(20)		
<1904		(100)	7		Leiden U.*
>1904		(100)	(7)		Leiden U.*
	Scandinavia				
1908	Uppsala U.*	327	21		
<1910		(200)	(15)		Copenhagen U. & Poly.
1910	Helsinki U.*	320	16	2	

TABLE D.1 *(continued)*

Date[a]	Wholly new quarters	Cost[b] 1000 marks	Floor[c] area $10^2 m^2$	Factor[d]	Enlargements, etc.
	Switzerland				
1890	Zürich ETH*	600	40	2	
1899		(100)			Geneva U.
	United States				
1894	Chicago U.*	950	40		
1897	Columbia U.	1500	22	2.3	
1900	Pennsylvania U.	650	17		
1902	Minnesota U.	315			
1903	Stanford U.	(300)	22	1.5	
1903	Nat. Bur. Standards*	[e]1500	60		
(1903)	Ohio State U.	300			
(1904)	Nebraska U.	315			
1905	Cornell U.	1430	100		
1905		150			Michigan U.
1909	Princeton U.*	1700	60	1.4	
(1909)	Illinois U.*	950	85		
1912	Yale U.	(1500)			
1912	Iowa U.*	630	40	1.5	
1913		600	(14)		Chicago U.

[a]Date is of completion of the laboratory. As in previous tables, parentheses signify our uncertainty.

[b]As "cost" we give the expenditure on the building per se, its installations, fixtures, furnishings, and scientific apparatus, or, if a multipurpose structure, the expenditure apportionable to the facilities for pure physics, exclusive of astronomy, mathematics, electrotechnics, etc. Where instruction in applied physics or electrotechnics was given under the aegis of the professor of experimental physics—as, typically, in the French universities lacking separate technical institutes and faculties—we have perforce included those facilities in our cost figures. When supplied in our sources, further outlay, if any, for the purchase of the ground is given in the notes, but not included in the national totals of Table I. Where our sources are silent on this point, as they usually are, we assume that the stated costs are exclusive of the value of the land. Often our sources do distinguish the cost of the building per se from that of its installations and furnishings, and also from that of its scientific apparatus. But as the line between building and fitting on the one hand and between installations and apparatus on the other is so variously and ambiguously drawn, we have suppressed those distinctions in the table and give separately in the notes only the reported expenditures upon scientific apparatus.

Where the data are wanting we have guessed at the "cost" of the institute or enlargements, exclusive of land and moveable scientific apparatus, on the basis of the reported or supposed size of the laboratory, or a rough estimate of physics' probable share of an outlay. These figures are in parentheses; they have been introduced chiefly in order to form the national totals of Tables I and II.

[c]Total area of all floors including corridors, walls, residences, etc., or, in buildings shared with other departments, of that portion devoted to pure physics.

[d]An entry in the "factor" column indicates that the building was shared by physics with one or more other departments. In such cases the entry in the "cost" column is that of the physical laboratories alone, which entry multiplied by this factor gives the total expenditure on the building and its installations. (Thus, where omitted, the "factor" is tacitly understood to be one.) The estimate of physics' share of the total is in nearly all cases our own; usually it is based on the relative floor area occupied by physics, a conservative estimator since the facilities for this science were more cost-intensive than those for any other. Cf. note.[19]

[e]These laboratories are "non-academic"; i.e., not integral with one of our set of institutions of higher education and therefore not included in the national totals, Tables I and II. In the notes we have included data and documentation on a few other laboratories falling outside our set of institutions or period.

*The sources cited below give relatively detailed descriptions of these laboratories, including at least partial plans in almost all cases. Earlier laboratories of which good descriptions are cited in the bibliography are Cambridge, Berlin, Königsberg, Strassburg, Tübingen, M.I.T.; see also Junk, "Phys. Institute" (1905). Apart from the countries included in Table D.1, detailed descriptions have been published of the new physical institutes at U. La Plata. Bose, *Phys. Zs., 12* (1911), 1230–43. Also T.H. Kiev. De-Metts (1903).

Austria. Brünn, T.H.: *Festschrift* (1899), pp. 113–14, plates IV–VIII. Innsbruck: Kalinka, *Österr. Forschungsinstitute* (1911), p. 73. Prague (total cost of institute constructed for physics, astronomy, and mathematics, inasmuch as mathematics' share devolved upon physics in 1908): Tschermak-Seysenegg (1919), pp. 11, 22–23. Vienna, T.H.: *150 Jahre* (1965), *1*, 74; *2*, 64–5; Austria, Min. f. Kultus u. Unterr., *Neubauten* (1913), pp. 45–46. Vienna, U. (inclusive of 130,000 marks to supplement the apparatus collections of the two experimental chairs): *ibid*, pp. 5–10; Austria, Reichsrat, Haus der Abgeord., *Sten. Protokolle*, 18. Sess. (1908); 20. Sess. (1910), p. 2528; Vienna, U., *Institute u. Einrichtungen* (1929), p. 64. Vienna, U., Inst. f. Radiumforschung (exclusive of ground [34,000 marks], but inclusive of "construction" [200,000 marks] and "wissenschaftliche Ausstattung" [185,000 marks]): Austria, Min. f. Kultus u. Unterr., *Neubauten* (1913), pp. 10–12; S. Meyer, *Jahrb. d. Radioaktivität, 17* (1920), 4–7; Kalinka, *Österr. Forschungsinstitute* (1911), p. 79.

Belgium. Brussels, U., *1884–1909*, p. 88.

British Empire. Aberdeen U.: *Handbook* (1906), pp. 84–88. Adelaide, U.: *Calendar* (1910). Belfast, U.: Moody and Beckett, *1* (1959), 433, 437. Birmingham, U.: G.B., Bd. of Ed., *Reports* (1909–10), p. 2. Bristol, U.: *ibid.*, p. 46. Cambridge U.: Cavendish Lab., *History* (1910), pp. 9–11; *Nature, 78* (1908), 152–53. Cardiff, U.: G.B., Parl., H. of C., *Sess. Papers, 19*, 1909, 778–79, 836; *Nature, 80* (1909), 127–28. Dalhousie, U.: *Cal-*

endar (1914-15). Dublin, Trinity Coll. (including 10,000 marks for instruments): *Calendar* (1911-12), p. 326; *History* (1947), pp. 60-62. Dundee, U.: G.B., Bd. of Ed., *Reports, 2* (1911-12), 434-35; *2* (1912-13), 463; *2* (1913-14), 498-500. Edinburgh, U.: *Calendar* (1907-08), pp. 653-57; *Nature, 75* (1906), 20-21. Glasgow, U.: G.B., Parl., H. of C., *Sess. Papers,* 1905, *61*, 304; 1906, *92*, 529; 1907, *65*, 640; 1908, *86*, 1914; 1909, *69*, 628; MacKie, *U. Glasgow* (1954), pp. 291-94, 310n. Leeds, U.: *Annual Report* (1899-1900), p. 33; Chapman, *U. Sheffield* (1955), p. 452; Shimmin, *U. Leeds* (1954), pp. 140-41; *Nature, 78* (1908), 257-58. Liverpool, U.: *Nature, 71* (1904), 63-65; G.B., Parl., H. of C., *Sess. Papers*, 1907, *22* (Cd. 3409), 225; 1908, *28* (Cd. 2789), 180; 1909, *19* (Cd. 4879), 236, 273. London, R. Coll. Sci.: *ibid.*, 1906, *31* (Cd. 2956), 565-71. London, R. Institution, Davy-Faraday Lab. ("£38,000 is sunk in the building and its equipment while the remaining £62,000 constitutes the endowment fund"): *Nature, 55* (1896), 208. London, U. Coll.: T.R. Smith, *Journ. R. Inst. Brit. Arch., 1* (1894), 286-87; *Building News, 68* (1895), 45; G.B., Bd. of Ed., *Reports* (1900-01), p. 254; Bellot, *U. Coll. Lond.* (1929), pp. 379, 390, 408; *Nature, 45* (1892), 348-49; *51* (1895), 298-99. Manchester, U.: *Phys. Labs.* (1906), pp. 127-29; Harker, *Nature, 76* (1907), 640-42; *Nature, 58* (1898), 621-22; *89*, (1912), 46; G.B., Bd. of Ed., *Reports* (1912-13), p. 262. Montreal, McGill U. (including 380,000 marks for moveable equipment): *Annual Report* (1894-95), p. 34; *Annual Calendar* (1899-1900), pp. 46-48, 153-55; *Formal Opening* (1893), pp. 63-67; Eve, *Rutherford* (1939), p. 64; Eve, *Nature, 74* (1906), 272-75; G.B., Parl., H. of C., *Sess. Papers,* 1906, *31* (Cd. 2956), 569. National Physical Laboratory (initial outlay of 410,000 marks for remodeling, installations, and equipment, plus 190,000 marks in 1905-06 for enlargements and new buildings): G.B., Parl., H. of C., *Sess. Pap.*, 1907, *68*, 367-80; *Nature, 63* (1901), 300-02; *65* (1902), 466-67, 487-90; *74* (1906), 205-06; *91* (1913), 464-65; Teddington, Eng., N.P.L., *Report* (1906), pp. 58-61 and following plans; *ibid.* (1907), pp. 85-97 and following plates and plans. Nottingham, U.: G.B., Bd. of Ed., *Reports* (1900-01), p. 364; *ibid.* (1899-1900), in G.B., Parl., H. of C., *Sess. Papers,* 1901 (Cd. 8165), p. 372. Oxford, U.: A. Chapman citing *Univ. Gazette*, 13 Oct. 1908, p. 66. St. Andrews, U.: Cant, *St. Andrews* (1946), p. 146. Sheffield, U.: G.B., Bd. of Ed., *Reports* (1900-01), pp. 375, 383-85; Chapman, *U. Sheffield* (1955), p. 452. Sydney, U.: *Calendar* (1901), p. 117; *Short Hist. Account* (1902), p. 104. Toronto, U.: Cooke, *Acad. Efficiency* (1910), p. 97; Langton, *McLennan* (1939), pp. 35-39; Loudon, *U. Toronto Monthly, 8* (1907), 42-47.

France. Clermont, U.: France, Min. instr. publ., *Enquêtes,* Nr. 104 (1910-11), p. 148. Gif, Lab. d'Essais radioact.: Danne, *Phys. Zs., 13* (1912), 565-74. Lille, U. (exclusive of value [c. 160,000 marks] of the land): Moissan and Matignon, *RGS, 6* (1895), 477-93; Nichols, *Phys. Rev., 3* (1896), 232. Marseilles, U.: France, Min. instr. publ., *Enquêtes*, Nr. 104 (1910-11), pp. 78-79; Nr. 108 (1912-13), pp. 118-19. Nancy, U. (exclusive of value [c. 80,000 marks] of the land): *Séance de rentrée* (1906), pp. 111-12; *U. Nancy 1572-1934,* pp. 88-93; France, Min. instr. publ., *Enquêtes,* Nr. 101 (1909-10), p. 299. Paris, Conservatoire nat. des arts et metiers, Lab. d'essais mécaniques, etc.: *Bulletin, 1* (1903-04); Teddington, Eng., Nat. Phys. Lab., *Report* (1903), p. 9, reports £27,000 for buildings, £20,000 for equipment, and £5,500 for the first year's budget. Paris, École mun. phys. et chim.: *Rapport Gen.* (1900), fig. 1 at end. Paris, U. (excl. of "Mecanique physique," 700m^2): France, Min. instr. publ., *Statistique, 3* (1889); Nenot, *Nouvelle Sorbonne* (1903); Berget, *La Nature, 26* (1898), 225–27; *Nature, 58* (1896), 12–13. Paris, U., Curie lab.: E. Curie, *Mme. Curie* (1937), pp. 236–37; Paris, Conseil U., *Rapport* (1908–09), p. 31. Paris, U., PCN (ten labs for elementary instruction in physics, chemistry, and biology): Paris, Conseil U., *Rapport* (1899–1900), p. xxxviii. Paris, U., Radium Inst. (exclusive of value [100,000 marks] of the land): E. Curie, *Mme. Curie* (1937), pp. 285–88; France, Min. instr. publ., *Enquêtes,* Nr. 101 (1909–10), p. 43. Poitiers, U.: *Institute of Physics* (1919); *Histoire* (1932), p. 417. Toulouse, U.: France, Min. instr. publ., *Statistique, 3* (1889), 384–88; *U. d. Toulouse* (1929), pp. 192–94; Toulouse, U., *Rapport* (1900–01), pp. 11, 140; (1903–04), pp. 6, 120; (1905–06), p. 109.

Germany. Aachen, T.H.: *1870–1970, 1*, 70; Damm, *T.H. Preussens* (1909), p. 167. Berlin, T.H.: *ibid*, p. 34; *ZBBV, 21* (1901), 230–31. Berlin, Handelshochschule: Handke, *Dtsch. Mech.-Ztg.* (1907), pp. 57–58. Berlin, Landwirtschaftliche Hochschule: Börnstein, *Phys. Zs., 12* (1911), 551–58; Dtsch. Phys. Ges., *Verhl., 13* (1911), 206–08. Berlin, Phys. Tech. Reichsanstalt (figure given is expenditure on buildings and equipment of Division I [Physics], exclusive of cost of land [500,000 marks]): Carhart, Smithsonian Inst., *Ann. Report* (1900), pp. 404, 407; Pfetsch, *Minerva, 8* (1970), 575. Berlin, U. (exclusive of alterations totaling 130,000 marks): *ZBBV, 15* (1895), 45 (40,000 marks); *28* (1908), 29 (66,000 marks); *33* (1913), 30 (25,000 marks). Bonn, U.: Kayser and Eversheim, *Phys. Zs., 14* (1913), 1001–08; *ZBBV, 33* (1913), 31; Konen in Bonn, U., *Gesch., 2* (1933), 352; (exclusive of prior alterations totaling circa 50,000 marks): *ZBBV, 16* (1896), 41; *18* (1898), 46; *21* (1901), 28; *25* (1905), 53. Breslau, U.: *ZBBV, 23* (1903), 145–46; *19* (1899), 46; *20* (1900), 34; *30* (1910), 41; *31* (1911), 42; Meyer, *Phys. Zs., 6* (1905), 195; Lummer in Breslau, U., *Festschr., 2* (1911), 446–47; Breslau, U., *Chronik* (1898–99), pp. 28–29; (1900–01), pp. 32–33; (1909–10), p. 52; (1910–11), p. 52; (exclusive of alterations in 1906 costing 28,000 marks): *ZBBV, 26* (1906), 41. Danzig, T.H. (exclusive of scientific instruments): *Festschr. zur Eröffnung* (1904), pp. 14–21. Darmstadt, T.H.: *Gebäude* (1895), pp. 24–33, 71–72, 84–92; Hesse, Landstände, 2. Kammer, *Verhl.,* 29. Landtag (1894–97), Beilagen-Bd. 1, Beilagen Nr. 150–51. Erlangen, U.: Wiedemann, *Phys. Inst. U. Erlangen* (1896), p. 5; Bavaria, Landtag, K. d. Abgeord., *Verhl.*, 31. Landtag (1892), Beilagen-Bd. 14, p. 297; Beilagen-Bd. 16, p. 49; Kolde, *U. Erlangen, 1810–1910*, p. 470; *Nature, 58* (1898), 621; Wiener, *Phys. Zs.,* 7 (1906), 6. Frankfurt a. M., Phys. Verein: *Neubau* (1908), p. 78; Hartmann, *Dtsch. Mech.-Ztg.* (1907), pp. 146–47. Freiburg i. Br., U.: Durm, *ZBBV, 13* (1893), 93–95; *Hochschul-Nachr., 11* (1900–01), 182. Giessen, U.: Wien, *Phys. Zs., 1* (1899), 155–60; Hesse, Landstände, 2. Kammer, *Verhl.*, 29. Landtag (1894–97), Beilagen-Bd. 1, Beilage 81, pp. 115–16; 30. Landtag (1897–1900), Beilagen-Bd. 4, Beilage 525 (specifies, i.a., 33,000 marks for apparatus); Wiener, *Phys. Zs.,* 7 (1906), 11; *Hochschul-Nachr., 10* (1900), 109. Göttingen, U.: *Phys. Inst.* (1906), Table I; *ZBBV, 24* (1904), 62; *25* (1905), 53; Riecke, *Phys. Zs., 6* (1905), 881–92; Klein, *Phys. Zs., 1* (1900), 143–45; Göttingen, U., *Chronik* (1901), p. 56; Wiener, *Phys. Zs.,* 7 (1906), 6. Greifswald, U.: *ZBBV, 11* (1891), 419–20; Greifswald, U., *Chronik* (1890–91), pp. 39–40; (1901–02), pp. 39–40; *Festchr., 2* (1956), 456–59; (exclusive of subsequent structural alterations and improvements in installations totaling 29,000 marks): *ZBBV, 18* (1898), 45; *21* (1901), 27; *28* (1908), 29; *29* (1909), 41; *31* (1911), 42; *32* (1912), 50. Halle, U., *Chronik* (1889-90), pp. 38-39; (1902-03), p. 44; Schrader, *Gesch. U. Halle, 2* (1896), 327-28, 548; *ZBBV, 9* (1889), 32; Gebhardt, *Wiss. Zs. U. Halle, 10* (1961), 856; (exclusive of improvements in electrical installations totaling 28,000 marks): *ZBBV, 22* (1902), 27 (8,000 marks); *29* (1909), 41 (20,000 marks); (and toilets): *ZBBV, 32* (1912), 50 (5,500 marks). Hamburg, Phys. Staatslab.: Voller in *Hamburg in naturw. Beziehung* (1901), pp. 207-09; v. Melle, *Dreissig Jahre, 1* (1923), 110, 116, 260, 538. Hannover, T.H.: *ZBBV, 15* (1895), 465; Hannover, T.H., *Festschrift* (1956), pp. 38-40; *100 Jahre* (1931), p. 39. Heidelberg, U.: Baden, Stände-Vers., 2. Kammer, *Verhl.*, Heft 490 ("Staatsvoranschlag 1910-1911") Hauptabt. III, Titel X, pp. 106-07; Heft 502 ("Staatsvoranschlag 1912-1913"), Hauptabt. III, Titel III, pp. 50-51; Heft 502 (1913), Plan 2; Ramsauer, *Elektrotech. Zs., 34* (1913), 1364-67; Lenard and Ramsauer, *ibid., 31* (1910), 1015-17; *33* (1912), 1103-05; *35* (1914), 1125-27. Jena, U.: *Geschichte, 2* (1962), 676; Lexis, *Univ., 1* (1904), 585; *Hochschul-Nachr., 10* (1900), 152-78. Kiel, U.: *ZBBV, 23* (1903), 158; *20* (1900), 34; *34* (1914), 35; Schmidt-Schönbeck, *Phys. Kieler U.* (1965), pp. 12, 70, 112, 122; Kiel, U., *Geschichte, 1,* Pt. 2 (1965), 143. Königsberg, U.: *ZBBV,* 7 (1887), 13-14; (excluding installations totaling 51,000 marks): *ibid., 26* (1906), 41; *28* (1908), 29; *34* (1914), 35. Leipzig, U. (exclusive of land costing 79,000 marks, inclusive of machines, instruments and apparatus costing 256,000 marks): Wiener in Leipzig, U., *Festschr., 4,* Pt. 2 (1909), 38-40, 52; Wiener, *Phys. Zs.,* 7 (1906), 1-14; (low-temperature physics laboratory): Lilienfeld, *Zs. f. kompr. u. flüss. Gase, 13* (1911), 165-80, 185-93. Marburg, U. (an enlargement for which 213,550 marks were budgeted in 1905 was apparently not executed; the new building, begun in 1912, was to have been completed in 1914 but was delayed by the war): Schulze in *U. Marburg 1527-1927*, pp. 759-62; Lexis, *Univ., 1* (1904), 441; *ZBBV, 9* (1889), 32; *25* (1905), 53; *32* (1912), 50; *33* (1913), 31; *34* (1914), 36. Münster i. W., U.: *ZBBV, 19* (1899), 46; *23* (1903), 144-45; *31* (1911), 42; *32* (1912), 51. Munich, T.H.: Bavaria, Landtag, K.d. Abgeord., *Verhl.*, 34. Landtag, Beilagen-Bd. 4, Budget 26 (for 1906-07), pp. 314-15. Munich, U. (exclusive of cost of the land [190,000 marks] and utilizing in part a wing of the main university building): *ibid.*, 31. Landtag (1892), Beilagen-Bd. 14, p. 296; Beilagen-Bd. 16, p. 48; Lommel, *Acad. Revue, 1* (1894-95), 261-65. Rostock, U.: *Jahresber., 6* (1911), 49-51; Becherer, *Wiss. Zs. U. Rostock, 16* (1967), 829. Stuttgart, T.H. (including 9,000 marks for instruments): Koch, *Phys. Zs., 12* (1911), 818-31. Tübingen, U.: *Deutsche Bauzeitung, 24* (1890), 213, 217; Württemberg, Landtag, K. d. Abgeord., *Verhl.*, 37. Landtag (1907), Beilagen-Bd. 1, Heft 15, pp. 16-17; Tübingen, U., *Institute* (1928), pp. 41-43.

Italy. Bologna, U.: *L'université* (1921), pp. 89-91; Bastiani, *Giornale del genio civile, 45* (1907), 667-92. Genoa, U., begun 1914, completed 1924: *Università di Genova* (1923), pp. 83-96, 121. Padua, U.: *Annuario* (1893-94), p. 7; *suoi istituti scientifici* (1900), p. 40. Pavia, U.: *Annuario* (1899-1900), pp. 147-48. Pisa, U.: *Annuario* (1899-1900), appendice, p. 125. Turin, U.: *R. Univ. di Torino* (1900), p. 44; *Annali* (1898), p. 87.

Netherlands. Amsterdam, U. ("Laboratorium Physica" built 1885-87 at cost of 170,000 marks): Junk, "Phys. Institute" (1905), p. 207; Amsterdam, U., *Gedenkboek 1632-1932*, pp. 224, 248. Delft, T.H. (institute for applied physics and electrotechnics, 7,000 m^2 floor area, built 1900-03): Delft, T.H., *Gedenkschrift 1842-1905*, Bijlage VIII, VIIIa. Groningen, U.: H. Haga, "Het Natuurk. Lab." (1914); *Acad. Groningana 1614-1914*, p. 220; *Univ. Groningana 1614-1964, 1,* 139. Leiden, U. (numerous small enlargements): Natuurk. Lab., *Gedenkboek* (1904), p. 15; (1922), pp. 74-75.

Scandinavia. Copenhagen, U. and Polytek. (numerous small enlargements): *Aarbog* (1908-09), p. 383; *Den Polytekniske Laereanstalt* (1910), pp. 9-13; Lundbye, *Polytek.* (1929), p. 365. Helsinki, U.: *Redogörelse* (1905-08), pp. 224-25; (1911-14), pp. 219, 232; Tallqvist, *Fys. inrättningarna* (1911). Stockholm, Tekniske Högskola (building for electrical engineering and physics constructed 1914-17, total floor area 2,500m^2, total cost 973,000 marks): *Skrifter* (1918), pp. 30, 69; Henriques, *Skildringar, 2* (1927), 495, 501. Uppsala, U.: *Handlingar betr. nybyggnad för den Fys. Inst.* (1903); *Aarsberettelse* (1910), p. 49; Schück, *Birka* (1910), pp. 49-50; *Uppsala Univ. 1872-1897, 2,* 151.

Switzerland. Geneva, U.: *Historique* (1914), pp. 3, 17, 20-23. Zürich, ETH (exclusive of cost [164,000 marks] of the ground): Bluntschli and Lasius, *Schw. Bauzeitung, 10* (1887); *ZBBV, 9* (1889), 135-37; Leobner, *Zs. V. d. I., 40* (1896), 746-48; Guye, *RGS, 8* (1897), 102-09; *Festschr., 1* (1905), 346; *2,* 336.

United States. Chicago, U.: Reeves, et al., *Univ. Plant* (1933), pp. 123-24, 137-38; Meyer, A., in U.S. Nat. Museum, *Report* (1902-03), pp. 495-96; Goodspeed, *U. Chicago* (1916), p. 237; *President's Report* (1903-04), pp. 237-38. Columbia U.: Meyer, A., *op. cit.,* p. 373; Cooke, *Acad. Efficiency* (1910), p. 97; *Report of President* (1898). Cornell U.: Howe and Grantham, "Phys. at Cornell," p. 6; Hewett, *Cornell, 1* (1905), 346-47; *2,* 158-62; *Register* (1906-07), p. 171. Dartmouth Coll.: Nichols, *Phys. Rev., 12* (1901), 366-71. Illinois, U.: Carman, *Brickbuilder, 20* (1911), 257-60; Nevins, *Illinois* (1917), p. 361. Iowa, U.: Stewart, *Contributions Phys. Lab. U. Iowa, 1,* no. 5 (1912), suppl., 8-15. Michigan, U.: U.S., Office of Ed., *Reports, 1* (1905-06), 449; Hindsdale, *Hist. U. Michigan* (1906), p. 366; Randall, *J. Opt. Soc. Am., 44* (1954), 98. Minnesota, U. (including 38,000 marks for apparatus): Erikson, "U. Minn. Dept. of Phys.," pp. 11, 20, 60. National Bureau of Standards (exclusive of 100,000 marks for the land, but inclusive of 170,000 marks for equipment): Rosa, *Science, 17* (1903), 129-40; Cochrane, *Hist. NBS* (1966), pp. 68-73. Nebraska, U.: U.S., Office of Ed., *Reports, 2* (1903-04), 1429. Ohio State U.: Cope, *Hist. of O.S.U., 1* (1920), 333-36, 369. Pennsylvania U. (original building 1700 m^2 valued at 350,000 marks plus 300,000 marks of equipment, enlarged by nearly 100 percent prior to 1914): *Provost's Report* (1900-01), pp. 45, 207; *Guide* (1904), pp. 16-17; Nitzsche, *U. of Penna. Guide* (1914), pp. 93-94. Princeton U.: *Treasurer's Report* (1908-09), pp. 76-78; McClenahan, *Science, 32* (1910), 291; Cooke, *Acad. Efficiency* (1910), p. 97. Stanford U.: *Directory* (1903), maps following p. 44; U.S., Office of Ed., *Reports, 2* (1901-02), 1351. Yale U. (the Sloanes gave $500,000 for building, equipment, *and* endowment): *President's Report* (1909-10), p. 33; Chittenden, *Sheffield, 2* (1928), 470. The U.S., Office of Ed., *Reports* for the years 1899-1903 list further physical laboratories constructed at the following institutions not in our set: Cincinnati U., DePauw U., Bucknell U., Washburn U., Wesleyan U., Louisiana State U., Washington U., and Lehigh U.

and mortar, laying out twice as many marks for this purpose as the Germans. Even allowing for the much higher costs in Britain,[5] it is evident from the relative rates of change of investment (8 percent versus 2-3 percent) that Britain was bidding fair to surpass the Germans in physics plant. The American investment, again allowing for the price differential, although less than the British, was substantially greater and increasing faster than the German. Apart from the new Sorbonne, France was already lagging in the 1890's and by 1914 had fallen seriously behind.[6] Like France, the smaller European countries had built their physical laboratories in the 1870's and 1880's. In Switzerland, for example, the mammoth physical-electrotechnical institute at the ETH, completed in 1890, was the last of a series which included physical institutes at the Universities of Basel, Berne, and Zürich. Italy continued building desultorily through our period, and Scandinavia became quite active again after 1905; but Switzerland and the Netherlands built no new physical institutes until after 1914.[7] Austria, which circa 1900 appeared "most unaccountably behindhand in the provision of properly designed physical institutes for its universities," and whose projected physical institute at the University of Vienna was admitted by the Minister of Education to be the "problem child among all the construction projects in our monarchy,"[8] did much to improve its situation in the decade before the war—in Vienna at any rate.

The financing of these new buildings followed the national patterns already revealed in the funding of laboratory expenditures. In Germany, France, and Italy, despite help from philanthropists and industrialists—and municipalities, especially in Italy and France—the main sums had to come

[5] In the 1880's and 1890's physical institutes in Germany and Switzerland cost 15-22 marks per cubic meter to build. *ZBBV, 7* (1887), 13-14; *11* (1891), 419-20; *23* (1903), 145, 158; Bluntschli and Lasius, *Schw. Bauzeitung, 10* (1887), 20. In the few cases where we have this figure for Britain and the United States it is two or more times higher: *Nature, 71* (1904), 63, for Liverpool; Reeves, et al., *Plant* (1933), pp. 123-24, for U. of Chicago. Cavendish Lab., *History* (1910), p. 11.

[6] For the construction program of the 1880's see France, Min. instr. pub., *Statistique, 3* (1889); Liard, *Enseignement supérieur, 2* (1894), 368 ff.

[7] The Netherlands built a large applied physics institute at the T.H. Delft, 1900-03, and new physical institutes at Leiden and Utrecht, 1915-23.

[8] Quotations of, respectively, Küchler, "Phys. Labs. Germany" (1906), p. 196, and Graf Stürgckh in Austria, Reichsrat, Haus d. Abgeord., *Sten. Prot.,* 20. Session (1910), p. 2527. Cf. the complaint of one representative, Daszynski, ("all universities in Austria suffer from a lack of teachers, from poor, inadequate budgets, and above all from miserable quarters") in *ibid.,* p. 2637; and, in general, *ibid.,* pp. 2087, 2475-87, 2526-44, 2601-05, 2627-49, 2685-717; Prokop, *Ausbau der T.H.* (1896), pp. 15-24; Innsbruck, U., *Bericht* (1905-06), pp. 7-9.

from the government.[8a] This was at least equally true in Austria, Switzerland, the Netherlands, and Scandinavia. In English-speaking countries private subscribers furnished much of the money, and in the United States in particular many of the largest physics institutes owed their existence and maintenance to rich individuals.[9]

As one would expect, physical institutes increased in average size and cost during the years 1890 to 1914. In Germany the trend is especially clear; the average cost of the eight wholly new institutes constructed in the decade 1890–99 was 250,000 marks, that of the six constructed in 1905–14 was 450,000 marks. The most powerful force behind the increasing size of physical institutes was the growth in the numbers of students in elementary physics courses. (In the German-language universities of Central Europe it was not uncommon that two or three times as many students registered for the lectures in experimental physics as could be seated in the lecture hall.[10]) Yet the data of Table D.1 suggest that the increase in the size of physical institutes was not so great as the increase in their cost: even though some of the more costly constructional requirements and installations of the 1890's were gradually discarded (see below, section D.3), the expenditure per square meter of floor space rose from about 100 marks in 1890–99 to about 150 marks in 1905–14. Half the difference may be ascribed to inflation in this period, but the remaining half is a measure of

[8a]An exception is the *consorzio* formed for building or extending science institutes at U. Bologna: 1.9 million marks from the state, 2.6 from the commune of Bologna, 1.3 from the city, 2.2 from the Cassa di risparmio. Italy, Min. pubbl. istr., *Monografie, 1* (1911), 28.

[9]Hewett, *Cornell Univ., 1* (1905), 346–47; *2,* 150–62; Goodspeed, *History of Univ. Chicago* (1916), pp. 179–93, 237–40, 428; French, *Johns Hopkins Univ.* (1946), p. 61 and passim; Hall in Morison, ed., *Development of Harvard Univ.* (1930), pp. 277–90; Princeton Univ., *Treasurer's Report* (1908–09), pp. 76–78; Yale Univ., *President's Report* (1909–10), p. 33. In the British Empire among the 17 wholly new quarters listed in Table D.1 at least 8 were financed largely or entirely by private philanthropy: McGill, Royal Institution, Manchester, Liverpool, Dublin, Glasgow, Oxford, and Dundee.

[10]The most notorious case was the University of Vienna, for which see sources of note 8, supra. At T.H. Darmstadt in 1896–97, 382 students registered for the experimental physics lecture in a hall seating 177; at U. Zürich circa 1913 there were 316 students and a lecture hall seating 125. Hesse, Landstände, 2. Kammer, *Verhl.,* 30. Ltg. (1897–1900), Beilagen-Bd. 1, Beilagen 15 and 70; Zürich (canton), Erziehungsrat, *U. Zürich* (1938), p. 813; likewise Basel, U., Thommen, *1884–1913,* pp. 116–18. The building of new lecture halls was thus a common form of enlargement of German physical institutes, e.g., T.H. Berlin (1897), U. Freiburg (1900), U. Tübingen (1908). Cf. Voigt, *Phys. Forschung* (1912), p. 16, complaining of the "overcrowding which, in accordance with our prediction, set in within a few years after the dedication [1905] of the Göttingen institute."

the increasing cost-intensiveness of the *Wissenschaftsbetrieb* in physics.[11] Or, to reverse the argument as the physicists of that generation liked to do, the increase in the number and expense of the instruments and installations requisite in a physical laboratory demanded a corresponding increase in the size of the institute if these facilities were to be efficiently used.[12]

Increasingly the object of these facilities was to aid research. "In previous decades," Conrad Dieterici explained to his education ministry in 1906 when assuming the chair at Rostock, "the task of physics extended essentially only to the confirmation of theoretical and mathematical speculations by clever and carefully thought out models. . . . Today the task of physics consists essentially in this: to pursue research."[13] Even more significant for laboratory design and expansion than the increasing research orientation of the teaching staff was the arrival of the research student. "It was not expected that students in the institute would carry out independent scientific investigations" when Tübingen's physical institute was erected in 1886–88. "In this respect," Paschen reported to *his* education ministry in 1906, "the situation has been essentially altered in the last ten years."[14] In the 1880's and 1890's it often happened that the research man was obliged regularly to remove his apparatus to make room for the students or the lecturer.[15] The necessity for such expedients diminished rapidly toward the turn of the century. At Leipzig, for example, the institute built in 1835 gave 12 percent of its work area to the laboratory; the institute of 1873, four times larger, gave 46 percent; and that of 1904, three times larger still, gave 60 percent, including many rooms for advanced and graduate students.[16]

The physics laboratory of the University of Illinois, completed in 1909,

[11] Cf. section C.3 above.

[12] W. Wien, *Aus der Welt der Wissenschaft,* p. 8.

[13] Dieterici quoted by Becherer (1967), p. 829. In that same year the apparatus collection of the University of Berlin physical institute was separated into two parts, one for lecture use and the other for scientific investigations. *Chronik* (1905–06), p. 171.

[14] Paschen to Naturwissenschaftliche Fakultät, 20 Feb. 1906 (U. Archiv, Tübingen, 117/904), published in Württemberg, Landtag, Kammer der Abgeord., *Verhl.,* 37. Ltg. (1907), Beilagenbd. 1, Heft 15, p. 16. The need to accommodate specialists (majors) was emphasized by K. Ångstrom in his justification of a new institute at Uppsala. *Handlingar* (1903), pp. 4, 17. Quantitative indices of this increase in the numbers of research students are given in Table A.3 and in the notes to Table A.4.

[15] Bellot, *Univ. College, London* (1929), p. 312; Rayleigh, *J.J. Thomson* (1942), p. 18; Schuster, *Biog. Fragments* (1932), pp. 56–57; *Nature, 65* (1902), 587–90; Heidelberg, U., *Geschichte* (1961), pp. 421–23; Langevin, "Phys. au Coll. de France" (1932), p. 74.

[16] Compiled from Wiener, Leipzig, U., *Festschrift, 4,* pt. 2 (1909).

may further illustrate the growing acknowledgment of the claims of research. The state legislature financed the expansion of the physics department primarily because of a great increase in the number of engineering students needing basic physics. But the arrangement of the new building was determined not only by undergraduate demand, but also by the needs of "a considerable and growing amount of graduate work and of investigation."[17]

2. Architecture, Layout, and Installations

A new physical laboratory was "built" by the incumbent of the chair, the intended director of the institute, after an inspection of the most recent and renowned examples in his country and often abroad. Seeking to surpass these models, he planned the number, nature, and arrangement of the rooms, their facilities and equipment, and the special constructional requirements to be met.[18] Along with government or university architects, he personally supervised the design, construction, and outfitting of the new laboratory to specifications more specialized and exacting than for any other type of scientific institute.[19] He was usually "guided by the principle that all architectonic elaboration was to be restricted as far as possible in favor of the greatest possible consideration of all practical requirements"; indeed, "more than one professor has suggested that the best style for a laboratory would be that of the common workshop, and perhaps with saw-tooth roof construction."[20] Although such utilitarianism often pre-

[17] U. Illinois, Physics Laboratory, *Contributions* (1912), pp. 6, 12.

[18] Junk, "Phys. Institute" (1905), pp. 213, 219, 221, 231; Durm, *ZBBV, 13* (1893), 93; Montreal, McGill U., *Formal Opening* (1893), p. 40. De-Metts' "Fizicheskie instituty" (1899) and Küchler's *Phys. Labs. in Germany* (1906) were formal reports of such study tours. The practice also held in the United States; e.g., A.P. Carman, *Brickbuilder, 20* (1911), 257: "The writer had the responsibility for making specifications for the design of a physical laboratory for the University of Illinois, and was in consultation with the architects and superintendents during the erection and equipping of the building." Cf. G.W. Stewart, *Contrib. Phys. Lab. U. Iowa, 1* (1912), suppl., p. 8, where the department is collectively responsible.

[19] This is implied by many continental writers; e.g., Bluntschli and Lasius, *Schweiz. Bauzeitung, 10* (1887), 9. Of the five natural scientific institutes completed at U. Strassburg in 1884, the cost per cubic meter was highest for the physical institute (22.5 marks), least for the chemical (20.5 marks). *Handbuch der Architektur*, Teil IV, Halbband 6, Heft 2aI, 2nd ed. (1905), p. 161.

[20] Riecke in Göttingen, U., *Phys. Inst.* (1906), p. 51, and Carman, *Brickbuilder, 20* (1911), 257. Likewise W.E. Ayrton to A.G. Webster, 11 March 1908 (AIP, New York). Junk, "Phys. Institute" (1905), p. 205, quotes "the previous director of the Würzburg institute" [Kohlrausch?]: "If the government would agree to such a thing, I would have the building treated as much like a barracks as possible." Cf. Kohlrausch's

vailed to a large extent inside the building, in most cases its external form was determined by other considerations: one thinks of the romanesque style of the McDonald Physics Building at McGill, chosen as "the type of stability and permanence," and of the terra cotta ornamentation, asked and paid for by the city of Bologna, "to give a marked local artistic character" to the facade of the new physics institute of its university.[21]

The interdependence of utility and ornamentation is obvious in the most salient feature of a laboratory built before the turn of the century, its tower. At the opening of our period, a tower was regarded as essential for experiments on free fall, pendula, extensibility, manometers, etc., as well as for atmospheric and meteorologic observations. It was seldom omitted from institutes completed before 1900, and the architect often welcomed this opportunity for "architectonic elaboration," as Nenot did in the case of the tower on the new Sorbonne.[22] But with the growing research emphasis and the shifts in research interest in the physical institutes after 1900, towers came to be "looked upon by a good many physicists as of largely traditional value."[23] In the first years of the century they were usually reduced or transformed into mere observation platforms on the roof, but with the decline of atmospheric physics even this vestige had disappeared from most institutes built after circa 1910.[24]

The building itself, nearly always of brick and massively constructed, contained a cellar (of which the floor was generally only about a meter below ground level), a "ground" floor, and one to four additional floors. Ceilings were about three and a half meters high, somewhat higher in the

---

chronicle of the construction of the Würzburg physical institute (1875), extracted by Reindl, *Math. u. Naturw. an der U. Würzburg* (1966), pp. 222-39, and O. Wiener, *Phys. Zs.*, 7 (1906), 8.

21 Cox in *Formal Opening* (1893), p. 41; Bastiani, *Giorn. gen. civile, 45* (1907), 669. To our knowledge no academic physical institute of this period was permitted saw-tooth roof construction; it was however employed in the buildings of the National Physical Laboratory completed in 1906-07. *Report* (1906), pp. 58-61 and following plans.

22 Nenot, *Nouvelle Sorbonne* (1895), p. 28.

23 Cooke, *Academic Efficiency* (1910), p. 39.

24 On the tower as a standard feature of the older laboratories: Junk, "Phys. Institute" (1905), pp. 197-98. It was, however, omitted from the Königsberg, Tübingen, and Freiburg institutes. Among the institutes constructed 1897-1907 the Giessen, Leipzig, Bologna, Edinburgh, and Uppsala institutes had towers, but the Manchester, Münster, Breslau, Kiel, and Göttingen institutes had only observational platforms on the roof. After 1907 the only institute we know to have included a tower was that at Helsinki.

ground floor, somewhat lower in the cellar and uppermost stories.[25] The largest and, in several respects, most important feature, the lecture hall for experimental physics, was generally situated in the upper regions of the building, or else in a separate annex, to maximize the window area and in particular to admit an extensive skylight in its ceiling.[26] The principal lecture halls constructed in the 1890's usually seated 100 to 150 auditors; those built after 1910 accommodated 200 to 400.[27] Even the smallest institute had a second lecture room, chiefly for theoretical physics, with a third to a sixth the number of seats of the main lecture hall. Adjoining the latter was a small preparation room and next to that a large room housing the collection of apparatus in glass-walled cabinets.[28]

The elementary laboratories were generally also in one of the upper stories, but not, however, when the professorial preoccupation with stable supports for instruments extended to the exercises of the beginning classes. The optical laboratories, and especially the photographic darkroom, sat high in the building, as did the living quarters of the assistants. The residence of the custodian-mechanic—only one, or at most two, of this class of employee lived in the institute—was invariably in the cellar, adjacent to the lavatory, coal bunker, furnace, machine shop, motor generators, and storage batteries.[29] In the larger institutes the cellar would also contain one or more constant-temperature research rooms, and often a diffraction

[25] Where physics was quartered in a comprehensive building, as was typically the case at the German Technischen Hochschulen, the ceilings often reached 5 meters. Brünn, T.H., *Festschr.* (1899), p. 113.

[26] Küchler, "Phys. Labs. in Germany" (1906), p. 198; for earlier practices see Junk, "Phys. Institute" (1905), pp. 181–93. The Darmstadt, Giessen, Breslau, Leipzig, and Princeton experimental physics lecture halls were placed at the top of the institute; those at the universities of Amsterdam (1887), Freiburg i. Br., Leipzig, Frankfurt, Illinois, and at T.H. Darmstadt are described in the sources cited in Table D.1 as having skylights.

[27] An exception was the Münster lecture hall, built for only 72 students in 1899. *ZBBV, 23* (1903), 144. Much data on the size of lecture halls underlies our estimates of numbers of students in Table A.4.

[28] Küchler, "Phys. Labs. in Germany" (1906), p. 197; Junk, "Phys. Institute" (1905), pp. 193–95.

[29] The power generating and converting machinery was sometimes housed in a separate one-story annex; e.g., at Münster, Kiel, Göttingen, and Cornell. In one, most exceptional, case the mechanical workshops were on the top (fourth) floor of the institute. Halle, U., *Chronik* (1889–90), p. 39. The foregoing rules were often violated when the building housed more than one department, as in the U. Vienna institute (1913).

grating mounting some five to ten meters in diameter.[30] The principal research rooms were placed on the ground floor, or possibly the "first" floor. This was the vertical distribution in general. The horizontal was more variable, but often strongly influenced by "the present [1898] tendency" to keep the tramping boots of the elementary student as distant as possible from the regions devoted to research.[31]

Residences for custodians and assistants occupied only 5-10 percent of the floor area of German institutes. However, when the building included a director's residence—as it did in perhaps a quarter of these institutes—an additional 15-20 percent of the floor area and of the expenditure was diverted from scientific facilities. Among the finest of these residences was that in the physical institute of the University of Leipzig, finished in 1904 to the specifications of the director, Otto Wiener. It had twelve large rooms, not counting cellars, a veranda, and a sickroom.[32]

As Table D.1 shows, the total floor area varied by a factor of about five between the smallest and largest institutes erected in any given country. Moreover, the number of rooms into which this space was divided varied by about the same factor, from fifteen to twenty in the smallest (e.g.,

[30] Large spectroscopic installations in the lowest and most stable regions were provided in the Manchester, Göttingen, U. Vienna, and Bonn physical institutes. The most elaborate provision for subterranean constant temperature rooms was at the ETH Zürich (1890). In European laboratories that placed some importance upon this facility, the effect continued to be achieved by thermal isolation; e.g., at the Bonn institute (1913). The U.S. laboratories, e.g., Princeton, introduced mechanical refrigeration, following the lead of the National Bureau of Standards (1902).

[31] A. Schuster in *Nature, 58* (1898), 621; also recommended by Junk, "Phys. Institute" (1905), p. 188. The classical instance of the separate entrance was Kundt's Strassburg institute (*Festschr. zur Einweihung der Neubauten* [1884], p. 61) as appears from Koch, *Phys. Zs., 12* (1911), 819, who speaks of "the usual division of the institute." The arrangement recurred at Tübingen (1888); at Freiburg (1890), where Emil Warburg placed "special importance" upon it (Durm, *ZBBV, 13* [1893], 94); and again at Darmstadt (1895), Bologna (1907), Frankfurt (1908), Illinois (1909), and Bonn (1913), all of which institutes had separate entrances to if not separate buildings for the main lecture hall.

[32] Wiener in Leipzig, U., *Festschrift, 4,* pt. 2 (1909), 38-40; cf. Meyer, *Phys. Zs., 6* (1905), 194-98, Wiedemann, *Phys. Inst. Erlangen* (1896), p. 8, and Kayser, "Erinnerungen" (1936), p. 255, on the parliamentary opposition to the provision of this facility for the director. The residence's roughly proportionate share of the expense is indicated by the data for Breslau in *ZBBV, 23* (1903), 145-46. The physical institute built at Freiburg i. Br., 1888-90, was exceptional in containing, as it appears, not even a custodial residence. Durm, *ibid., 13* (1893), 93-95. The free living accommodations for custodians and mechanics and for assistants help account for the relatively low total of German wages (Table C.2) and low salaries (Table B.1), respectively.

Münster) to nearly one hundred in some of the largest (e.g., Leipzig). This rather surprising proportionality, deriving, to quote the dissenting Schuster once again, from "the practice now in fashion which favors small rooms for single students,"[33] gradually failed in the first decade of the new century owing to a shift to larger, more communal and less specialized apartments. (Thus where the Leipzig institute averaged about 70 m^2 per room, the Bonn institute, although smaller, averaged about 80 m^2, and the larger Princeton institute about 110 m^2.) Even before the opening of our period at least one institute (Königsberg, 1887) included an elevator for the transport of heavy apparatus and of personnel—for scientific purposes.[34] From 1900 onward elevators seem to have been a standard facility in the larger institutes.[35]

Descriptions of physical institutes built from the mid-1890's onward are generally distinguished by the attention and detail they devote to the facilities for the production and distribution of electric currents of various types, strengths, potentials, etc.[36] The prime mover in the European physical institutes built in the 1880's and 1890's was a "gas engine"; i.e., an internal combustion engine drawing its fuel not from a gasoline tank but from the municipal gas mains.[37] This motor drove one or several electrical generators supplying (relatively little used) alternating current and, more important, rectified current to charge the tiers of storage batteries that constituted the principal sources of electric power. Within a very few years after the turn of the century nearly all institutes switched from the municipal gas mains to the newly installed municipal electric mains, whose currents, direct in nearly every case, were again used to charge the accumulators.[38] The number of elements (at 20-60 marks each) in these batteries and their capacities increased rapidly with time as well as

[33] *Nature, 58* (1898), 621; again Schuster, *Progress of Physics 1875-1908* (1911), p. 18. This arrangement apparently derived from the U. Berlin institute (1878). Guttstadt, *Naturwiss. Staatsanstalten Berlins* (1886), pp. 135-48; Kayser, "Erinnerungen" (1936), pp. 105-07. It found its most extreme expression in the French physical institutes of the 1890's, Lille and the new Sorbonne.

[34] *ZBBV, 7* (1887), 13-14.

[35] McGill, Manchester, Liverpool, Bologna, Breslau, Leipzig, Göttingen, and Bonn physical institutes are all credited with elevators by the sources cited for Table D.1.

[36] Extensive detail appears in the description of the Göttingen institute (1905), and also in descriptions of the Darmstadt, Giessen, Leipzig, Stuttgart, Helsinki, and Bonn institutes cited in notes to Table D.1.

[37] Nichols, *Phys. Rev., 3* (1896), 232-33. In 1902 the physics department of Clark U. converted from a steam engine to an "oil" (diesel?) engine as prime mover. A.G. Webster in *Report of the Pres.* (1902), p. 35.

[38] Küchler, "Phys. Labs. in Germany" (1906), p. 202. The U. Berlin institute hooked up to the AEG central station in 1890. *ZBBV, 10* (1890), 35. The U. Basel institute

with the size of the institute. The frequency and voltage range of the available alternating currents depended on these same variables.[39]

From these sources several circuits of varying current and voltage capacity led up to the laboratories, and complex switchboards allowed an experimenter to literally call for the galvanism of his choice. Here too the technology was advancing very rapidly, and many an institute built in the early 1890's had to rewire before the end of our period.[40] Most of this galvanic diversity was at least nominally in the service of instruction. Moreover the experimental table in the main lecture hall, in addition to a variety of other special utilities, was supplied by special conduits with exceedingly heavy currents of one, two, or even four thousand amperes.[41]

3. The Principal Desideratum: Undisturbed Measurement

In reporting the numbers, sizes, layouts, and outfittings of physical laboratories constructed in the generation before the war, we have barely touched upon the consideration that preoccupied their planners at the beginning of this period and that, though losing some of its urgency, re-

---

hooked up to the municipal system in 1901. Basel, U., *1884–1913,* p. 117. Thereafter: Halle in 1902. *Chronik* (1901–02), p. 43. Tübingen c. 1902. Kiel c. 1904. *ZBBV, 23* (1903), p. 158. Breslau in 1905. *Chronik* (1905–06), p. 41. Columbia in 1903. *Columbia U. Quarterly, 5* (1902–03), 295. Edinburgh in 1907. *Nature, 75* (1906), 20. Erlangen c. 1910. Bavaria, Landtag, K. d. Abgeord., *Verhl.*, 35. Ltg., Beilagenbd. XII, Budget 28, p. 154. In a few instances institutes built at a later date opted for their own power supplies rather than rely on the municipal electric systems. London, Roy. Coll. of Sci., *Calendar* (1911–12), p. 51; Kayser and Eversheim, "Bonn," *Phys. Zs., 14* (1913), 1006.

[39] In 1900 the Giessen institute had 45 elements; in 1905 the Göttingen institute had 100, each of about 100 ampere-hours capacity; in 1913 the comparable Bonn institute had nearly 200 elements of 150 to 400 ampere-hours capacity. Darmstadt, T.H., *Die neuen Gebäude* (1895), p. 91; Göttingen, U., *Phys. Inst.* (1906), p. 54; Wien, *Phys. Zs., 1* (1900), 157; Meyer, *ibid., 6* (1905), 196; Kayser and Eversheim, *ibid., 14* (1913), 1007. The Helsinki institute was wired to take 2,000 amps from accumulators, of which however it had acquired only 80, delivering a maximum of 576 amps, at its opening in 1910. Tallqvist, *Den nya byggnaden* (1911), p. 24.

[40] The McGill institute, opened in 1893, was entirely rewired in 1901, for "the wiring of the building for light and power, though the best of its kind when installed ten years ago, had become obsolete." John Cox in McGill U., *Annual Report* (1900–01), p. 31. See note to Table D.1 re U. Halle.

[41] This record was held by the Royal College of Science, according to its *Calendar* (1911–12), p. 51. On the elaborate outfitting of the experimental lecture theatres see: Küchler, *Phys. Labs. in Germany* (1906), pp. 199–200; Junk, "Phys. Institute" (1905), pp. 181–93; Göttingen, U., *Phys. Inst.* (1906), pp. 60–63; Leipzig, U., *Festschr.* (1909), pp. 42–44; Danzig, T.H., *Festschr. zur Eröffnung* (1904), pp. 20, 24; Koch, "Stuttgart," *Phys. Zs., 12* (1911), 826–29; Robins, *Technical School and College Building* (1887), pp. 143–44; McGill U., *Formal Opening* (1893), pp. 63–64; Tallqvist, *Den nya byggnaden* (1911), pp. 28–29.

mained to the end the most important single consideration in siting a physical laboratory: minimization of external interferences with the process of measurement. This desideratum and the provisions made for achieving freedom from particular classes of disturbing influences figure prominently in nearly every description of a physical institute erected in this period.[42] Among these "freedoms," three are especially characteristic of the epoch and of its changing objects and conditions of physical research. Most dearly and constantly valued was freedom from shaking or vibration.[43] Subordinate to it, but only just so circa 1890, was freedom from iron.[44] After the turn of the century this latter desideratum was largely superseded by freedom from electric trams.

Freedom from the disturbing influence of iron upon magnetic measurements appears to have reached its acme as a constructional principle just at the opening of our period. Although the avoidance of iron girders in the building as a whole, and the provision of at least one room without any iron fittings (pipes, radiators, hinges, nails) whatsoever, remained standard to about 1905,[45] in the early 1890's the condition of "complete freedom from iron" was imposed upon entire wings and even upon the entire building.[46] The reorientation of physical research in the middle of our period

[42]Junk, whose views are those of the 1880's, lists four "conditions" to be fulfilled in building a physical institute: freedom from shaking; freedom from air pollution; freedom from magnetic perturbations; freedom from temperature influences. "Phys. Institute" (1905), pp. 167–68. Similarly Bluntschli and Lasius, *Schw. Bauzeitung, 10* (1887), 9. To a later generation the provisions made for realizing these freedoms—in particular, freedom from shaking and from magnetic perturbations—appeared not antiquated (as contemporaries had fully expected) but merely ridiculous. Casimir, *Lecture* (1962), p. 14.

[43]The description of the U. Greifswald institute, e.g., begins: "The most essential requirement which must be made of the site of a physical institute is that its *Erschütterungsfreiheit* be as complete as possible." *ZBBV, 11* (1891), 419. For like, but less categorical, considerations at the end of our period: Tallqvist, *Den nya byggnaden* (1911), p. 7; Kayser and Eversheim, *Phys. Zs., 14* (1913), 1001.

[44]H. Haga's desiderata as stated in 1892 at the inauguration of his new institute: "1. facilities for setting up instruments steadily, and 2. freedom from iron." *Acad. Groningana 1614–1914*, p. 486.

[45]E.g., Königsberg, *ZBBV*, 7 (1887), 14; Greifswald, *ZBBV, 11* (1891), 419; Darmstadt, T.H., *Die neuen Gebäude* (1895), pp. 85–86; Münster and Breslau, *ZBBV, 23* (1903), 145–46; Göttingen, U., *Phys. Inst.* (1906), p. 51; Uppsala, U., *Handlingar* (1903), p. 14; Durham College of Science (U. Newcastle) (1888); and McGill, U., *Formal Opening* (1893), p. 63.

[46]Haga's Groningen institute (1892) was completely iron-free. *Acad. Groningana 1614–1914*, pp. 488–89. More than half of Warburg's Freiburg institute (1890) was subjected to this condition. Durm, *ZBBV, 13* (1893), 94. Halle and Giessen had iron-free wings. Halle, U., *Chronik* (1889–90), p. 38; Wien, *Phys. Zs., 1* (1900), 155–56. In this respect Braun's Tübingen institute (1888) was extraordinarily mod-

weakened this imperative so quickly and thoroughly that after 1905 steel girders appear promiscuously and nothing more is heard of freedom from iron.[47]

The concern for freedom from vibration, on the contrary, remained alive throughout our period, although by 1914 it had lost the primacy which it had enjoyed a generation earlier. "The great desideratum of a physical laboratory is a steady working-table, and this," as a British author observed in 1887, "is difficult to secure."[48] In general this object was sought in two ways: first, by founding and constructing the building as a whole as solidly as possible; second, by provision of special piers, isolated from the building, upon which instruments could be mounted. The precedents for both these proceedings derive from an earlier period; indeed, in the mid-1870's 310,000 marks were spent on the foundations of the University of Berlin physical institute, which was also equipped with many free-standing piers.[49] The attention to foundations continued undiminished, although Berlin's record remained unequaled. In 1900 Saxony was prepared to spend 100,000 marks digging an isolation trench four meters deep around the foundations of the Leipzig institute; in 1912 Baden appropriated an extra 90,000 marks to dig the foundations of the Heidelberg institute down to bedrock.[50]

The isolated pier was considered an indispensable fixture at the opening of our period.[51] Stone or slate ledges, brackets, and tables mortared into

---

erate, having no iron-free wing or even rooms, but avoiding *hidden* iron masses in the walls, etc. Tübingen, U., *Festgabe* (1889), p. 5; *Deutsche Bauzeitung, 24* (1890), 213.

[47]Nagaoka to Rutherford, 22 Feb. 1911: "The famous magnetic observatory without iron built by Kohlrausch is now rotten; more important works on vacuum discharge have absorbed the attention of Würzburg physicists." Badash, *Phys. Today*, April 1967, pp. 55-60. Steel girder construction is admitted in the descriptions of the Liverpool (1904), Edinburgh (1907), Princeton (1909), Stuttgart (1909), Iowa (1912), and U. Vienna (1913) physical institutes cited in the notes to Table D.1. However, roughly half the Imperial College (1907) physical laboratory "was constructed as far as possible without iron in order to minimize magnetic disturbances." *Calendar* (1911-12), p. 50.

[48]Robins, *Tech. School and College Building* (1887), p. 116, who was not, of course, referring to any deficiency of British carpentry. Also for anti-vibration measures of the 1880's, see Junk, "Phys. Institute" (1905), pp. 169-71.

[49]Guttstadt, *Naturwiss. Staatsanstalten Berlins* (1886), pp. 135-39, an extraction of the account by Kleinwächter, *ZBBV, 1* (1881), 359-61.

[50]Wiener, *Phys. Zs.*, 7 (1906), 7-8; Baden, Landtag, Zweite Kammer, *Staatsvoranschlag 1912-13*, Hauptabt. III, Titel III, pp. 50-51.

[51]O.E. Meyer complained of the inadequacy of his institute dating from 1867:

the walls existed in all rooms for research and in many for instruction, but in the thirty years before 1900 they were regarded as insufficient protection against vibrations originating within and without the building. It was agreed that the isolated pier gave the requisite stability, and physicists of that generation—such as Helmholtz, Ayrton, and O. Wiener—made extensive efforts to test different modes of construction, isolation, etc.[52] Generally the institutes of the late 1880's and early 1890's had at least half a dozen brick or masonry piers, roughly half a meter square, projecting from the foundations of the building into the laboratories in the ground floor, plus one to several especially massive piers on completely independent foundations.[53] But as research interests and instrumentation shifted, and experience insisted that isolated piers were not outstandingly effective, the consensus began to shift in the late 1890's.[54] After 1900 the

---

"Still worse is that solidly founded piers for mounting of measuring instruments are completely absent." Breslau, U., *Chronik* (1892-93), p. 29. But when he did finally get a new institute in 1900 the Prussian education ministry allowed him, "unfortunately," only one independently founded pier. Meyer, *Phys. Zs., 6* (1905), 196. Here once again Braun in Tübingen (1888) and Haga in Groningen (1892) present, respectively, moderate and extreme solutions. The latter had a most elaborate system of separately founded pillars and piers; the former incorporated none, but, with great success, simply put half the ground floor directly on the ground and elsewhere relied on massive arched beams. *Deutsche Bauzeitung, 24* (1890), 213; *Acad. Groningana 1614-1914*, pp. 486-88.

[52]Guttstadt, *Naturw. Staatsanst. Berlins* (1886), pp. 135-36; Robins, *Techn. Bldg.* (1887), p. 116; Wiener, *Phys. Zs.,* 7 (1906), 7-8.

[53]Strassburg, a model in this respect as well, had three independently founded piers and sixteen resting on the cellar beams. *Festschr.* (1884), pp. 63-66. Harvard's Jefferson Physical Laboratory (1884) boasted "a large rectangular tower standing on an independent foundation." *Guide Book* (1898), pp. 83-86. Others were: Ayrton's Yedo College of Engineering. Robins, *Techn. Bldg.* (1887), pp. 145-46. Freiburg (1890). Durm, *ZBBV, 13* (1893), 94. Halle (1890). *Chronik* (1889-90), p. 38. Greifswald (1891). *ZBBV, 11* (1891), 420. McGill (1893). *Formal Opening*, p. 65. Darmstadt (1895). *Die neuen Gebäude*, p. 91. Poitiers (1894) had one independently founded pier. *Institute of Physics* (1919). U. Paris, Lippmann's laboratory. Nenot, *Nouvelle Sorbonne* (1895), p. 28. (The French congratulated themselves for obviating "the inconvenience pointed out by the German professors" that currents of "disagreeable" air rise from the cellar along these isolated piers.) Aberdeen (1898). *Handbook to City and Univ.* (1906), p. 86. The London tradition deriving from Ayrton and Cary Foster was peculiar; it demanded, even for the elementary student, a work table whose legs stood upon brick supports independent of the floor: Robins, *Techn. Bldg.* (1887), pp. 116, 145-46; Smith, *J. Roy. Inst. Brit. Arch., 1* (1894), 287; Imperial Coll. of Sci. and Tech., *Calendar* (1911-12), p. 50.

[54]Küchler, "Phys. Labs. in Germany" (1906), p. 197. Although Breslau's institute, completed in 1900, had one free pier, those at Münster (1899) and Kiel (1901) appear to have had none. *ZBBV, 23* (1903), 144-46, 157-58. Leipzig (1904) certainly

construction of a physical institute with these encumbering piers was exceptional.[55]

The chief external source of mechanical vibration was nearby vehicles, including the greatest of all menaces, the new electric tram. The real threat of the electric tram, however, was an entirely new sort of physical disturbance: variable electromagnetic fields arising from intense transient currents. Shortly after the new physical institute of Turin was completed in 1898, a tram line was laid only a few meters from it. To dramatize the unhappy situation, the institute director wrote that "every investigation requiring even moderate precision has become impossible, and the institute has been reduced to such a condition as no longer to deserve its name."[56] In Jena the physical institute built in 1884 and enlarged in 1894 was considered to have been rendered untenable by the construction of an electric tram line some fifty meters distant, and a new institute was begun in 1900.[57] Various countermeasures and recourses short of complete relocation were employed: interdicting the use of the steel rails as return con-

---

did not. Wiener, *Phys. Zs., 7* (1906), 7-8. Nor did Göttingen. Gött., U., *Phys. Inst.* (1906), p. 51. Nor did U. Iowa (1912). Stewart, *Contribs.* (1912), p. 19. Stuttgart and Princeton (1909) both relied solely upon making the building itself as massive and rigid as possible; the latter with complete success, the former with very little. Koch, *Phys. Zs., 12* (1911), 812-13; McClenahan, *Science, 32* (1910), 292. The Bonn (1913) institute had three cellar rooms with separate foundations, but no isolated piers. Kayser and Eversheim, *Phys. Zs., 14* (1913), 1003.

[55]The exceptions known to us are U. Illinois (1909) and Righi's U. Bologna (1907). Carman, *Brickbuilder, 20* (1911), 260; Bastiani, *Giorn. genio civile, 45* (1907), 678. Such white elephants were still occasionally constructed as late as the 1950's and soon afterwards recognized as "probably one of our mistakes." London, Inst. of Phys., *Design Phys. Labs.* (1959), p. 31. Yet they continue to be recommended by British experts. Steffens in *The Design of Physics Buildings* (1969), p. 25.

[56]Turin, U., *R. Univ. di Torino* (1900), p. 44. Likewise Rome, U., *Annuario* (1899-1900), Appendice, p. 15: "The physics institute would answer every scientific desideratum if it did not have to fight against the relative slimness of its annual budget, and if it were not disturbed and menaced by the electric tramways which make high precision electrical measurements impossible." Among the institutes seriously hurt by electric streetcars were those at U. Berlin; U. Breslau (*Chronik* [1892-93], pp. 29-31 and O.E. Meyer and K. Mützel, *Elektrotech. Zs., 15* [1894], 33-35); U. Halle; U. Jena; T.H. Karlsruhe; U. Kiel (*Chronik* [1899-1900], pp. 45-46, and again [1904-05], p. 49); U. Basel (*Gesch. der Kollegiengebäude* [1939], pp. 49-50); Royal Coll. of Sci., London; Columbia U. (*Quarterly, 5* [1903], 296).

[57]Lexis, *Univ., 1* (1904), 585. McGill was able to avert such a "calamity" to their new physics building "by timely representations to the [traction] Company," presumably by the donor himself, McDonald. *Annual Report* (1895), p. 25. The avoidance of the streetcar threat became a primary consideration in the siting of new institutes. Rosa, *Science, 17* (1903), 130; Schulze in *Marburg 1527-1927*, p. 761.

ductors; acquiring magnetically shielded galvanometers; shifting research fields, e.g., to optics and spectroscopy. In some cases the financial means for retooling were provided by the traction companies as compensation, but only after protracted negotiation.[58]

[58]Halle's physical institute received 14,000 marks. *Chronik* (1902-03), pp. 44-45. T.H. Karlsruhe's was awarded 60,000 marks in 1901 after ten years of negotiation. Lehmann, *Gesch. Phys. Inst. Karlsruhe* (1911), pp. 80-81. In the same year the Royal College of Science, London, was promised 200,000 marks as compensation for the removal of the Kew observatory. G.B., Bd. of Ed., *Report for 1901 . . . on the Museums*, p. 56.

E. PRODUCTIVITY

In the preceding sections we have estimated for each of the leading scientific nations the number of academic physicists and the amounts of money spent upon them and their facilities. These monies supported all aspects of their activity: their private lives, their teaching, their research. The overriding importance of this last aspect of their total activity motivates our study. To conclude it we turn to the connection between the amounts of money spent upon academic physicists and the amounts of research they performed.

1. Output of Physics Papers

As measure of research done we take the volume of publication of physics papers in scientific journals.[1] Table E.1 contains the gross national products of physics papers, and their rates of change (papers per year), in six leading nations as given by four different estimators, two British and two German. The different measures of GNP show a very considerable agreement, and we therefore have some confidence in our best guess, column 5. Although our measure of volume of publication is number of papers, we give also number of pages in those papers. The ratio of these two quantities has nearly the same value for the four principal contributors: for this group the number of papers is a good measure of volume of research publication. Moreover we find that no appreciable error is introduced by crediting each leading country with *all* the papers appearing in its core physics journals, regardless of the nationality of the author or the location of his laboratory.[2] Neither of these propositions holds for the smaller countries, whose journals allowed more space per article than did the core journals, and carried only a fraction of the research papers published by their physicists.

Table E.2 reduces our best guess of the GNP of physics papers, and its rate of change, to relative values, and includes four further estimators of the relative rates of production of physics research by the principal contributing nations. Once again the estimates agree remarkably. As expected Germany is far ahead; the British Empire is second, followed by a faltering France; the United States is a distant, but rapidly gaining, fourth, Italy a poor and declining fifth. The French, unwarrantably complacent about the state of science in their country, and that of physics in particular, ne-

[1] Sarell, *Productivity in 19th-Century Phys. Sci.* (1971).

[2] See Table A.8, from which it might appear that this procedure does a considerable injustice to the Americans. However, Table E.2 shows that this is not the case.

TABLE E.1

Gross National Products of Physics Papers, circa 1900

	Index of Royal Soc. Catalogue of Scientific Liter.[a]		Science Abstracts A (1903/04)[b]		Fortschritte der Physik[c]			Annalen Bei-blätter[d]	Best guess	
	Papers	Rate of change	Papers	Rate of change	Papers	Rate of change	Ratio of pages to papers	Papers	Papers	Rate of change
British Empire	530	22.0	440	50	320	5.3	8.6	410	420	45
France	290	2.1	360	37	380	-8.6	8.8	375	360	25
Germany	580	24.0	480	59	570	28.0	8.8	905	580	60
Italy	110	-4.6	130	3	120	-4.8	10.9	115	120	0
Netherlands	75	6.6	35	7	140	-7.0	14.7	45	80	5
United States	120	3.3	250	63	250	-3.7	9.4	190	240	55
All Other	210	4.6	190	11	240	-5.2		150	200	10
TOTAL	1900	58.0	1900	230	2000	4.0	9.5	2200	2000	200

[a]The number of papers produced in 1900 and the annual rates of increase (decrease) are derived from a 12 percent sample from the Royal Society London, *Cat., Subject Index, 3: Physics,* for 1900, and for 1895 and 1890. We have not included counts from the continuation of this work, the *International Catalogue of Scientific Literature,* as the distribution of publications by nation is quite evidently unrepresentative—presumably because the references to the literature of each country were supplied by national bureaus with highly diverse criteria of inclusion and energy in collection.

[b]Figures derived from *Science Abstracts,* Sect. A: *Physics, 6* (1903), 7 (1904), and *15* (1912), *16* (1913), 30 percent and 15 percent samples, respectively. We have entered 1903/4 because this was the first year in which physics was separated from electrical engineering and because the total number of papers agrees with our other sources for 1900.

[c]The number of papers and pages, and the rate of change of the number of papers are derived from c. 20 percent samples from the *Fortschritte der Physik, 56* (1900), *57* (1901), and *67* (1911), *68* (1912). Time series of the *subject* distribution of papers abstracted in the *Fortschritte,* 1890–1914, are given by Hirosige and Nisio, *Jap. Stud. Hist. Sci.,* 7 (1968), 93–113.

[d]The number of papers is derived from counts of the first four of the monthly lists of "current contents" of scientific journals published by the *Beiblätter zu den Annalen der Physik, 24* (1900).

TABLE E.2

Relative National Products of Physics Papers, circa 1900
(with rate of change per annum, Δ)

	"Best guess" (Table E.1)		Papers in principal nat. jrnls.[a]	Authors on relativity[b]	Authors of publications on spectroscopy[c]		Discoveries in physics[d]	
	Percent	Δ	Percent	Percent	Percent	Δ	Percent	Δ
British Empire	21	22	19	15	21	11	14(14)	13
France	18	13	10	13	13	22	13(16)	6
Germany	29	30	33	30	43	32	42(38)	50
Italy	6	0	15	5.5	3	4		
United States	12	27	9	11	10	20		
All others	14	8	–	25	10	13		
TOTAL	100	100	86	99.5	100	102	69(68)	69

[a]Germany: *Annalen der Physik;* France: *Journal de physique;* Britain: *Phil. Mag.;* United States: *Phys. Rev.;* Italy: *Nuovo Cimento.*

[b]Distribution by nationality of 1175 authors of publications on relativity through 1922. Lecat, *Bibliog. de la relativité* (1924), p. 201.

[c]Distribution by nationality of authors of publications on spectroscopy c. 1900. Kayser, *Phys. Zs., 39* (1938), 466–68.

[d]The data are based upon Rainoff, *Isis, 12* (1929), 287–319. The first figure on the left is calculated from Rainoff's original data for the period 1896–1900, the second from Rainoff's curve fitted to the data for the entire century. The figure on the right is the relative rate of increase between 1891–1895 and 1896–1900, computed from the fitted curve. Figures have been normalized to the sum of the "best guess" shares of Britain, France, and Germany in publication, column 1.

glected to arrest their relative decline, while the Germans, although well aware that they led, bore the challengers ever in mind.[3]

Interesting and suggestive as these figures may be, a meaningful comparison of them with our data on numbers and funding of academic physicists cannot be made until they have been reduced to that proportion of the papers in each country attributable to the class of physicists we have studied. For this purpose we have determined the fraction of papers in each of the principal national physical journals circa 1900 due to domestic contributors who were then affiliated, either as staff or students, with one of our set of institutions of higher education of that nation (Table E.3, first column; cf. Table A.8). The determination assumes that a time average expressing the difference in rates of publication between academic and nonacademic physicists may be replaced by a space average over the core journals at a given instant;[4] it further assumes that publication in the core journals by holders of nonphysics posts at institutions included in our survey is compensated by publication in nonphysics journals by their colleagues in physics. On this basis we give in Table E.3 the productivities of the academic physicists in the principal contributing nations c. 1900, and also those of a few individual laboratories.

In general the productivity indices for Germany, the Netherlands, France, and Britain assume remarkably similar values; nonetheless a rank ordering is fairly clear in both the indices of output per man and per mark. Italy, whose indices are only two thirds to one half those of this Northern European group, still outstrips the United States, even though we have boosted the American "per mark" indices by 50 percent to take account of the lower research purchasing power of the mark in the United States. The indices for the individual institutes are not very reliable; we give them nonetheless because of the particular interest they must have for the historian of physics.[5]

[3]E.g., Deslandres, Soc. Fran. de Phys., *Bulletin, 37* (1909), 3*. Outsiders saw clearly the imminent French decline; as Henry Crew noted in his diary, Paris, 25-26 July 1895, "the young men of France are, so far as I can see, doing very little in Physical research." By contrast the German physicists never ceased to complain—even at the dedication ceremonies of a new laboratory—of the niggardliness of their education ministries. Voigt in Göttingen, U., *Phys. Inst.* (1906), pp. 41-43, 197-99. Cf. Harnack, "Denkschrift (1909)," in Max-Planck Ges., *50 Jahre* (1961), pp. 80-94.

[4]Were it possible to take as the "given instant" a time interval sufficiently short that the journal in question only rarely included more than one paper by any given author, it would suffice to count contributors rather than contributions.

[5]Tables A.4 and C.2 permit computation of such indices for any laboratory for which the rate of publication is known.

TABLE E.3

Productivities of Academic Physicists, circa 1900

	Annual output of papers[a]		Papers per physicist[b]		Papers per 10^4 marks of:[c]	
	Percent of GNP	Number	Post-holding	All affiliated	Lab. expenditure	Total expenditure
British Empire	69	290	2.2	1.7	9.2	1.4
Cambridge U.[d]		25	2.5	0.9	10	
Manchester U.[e]		10	1.4	0.9	14	
Toronto U.[f]		2	0.7	0.4		
France	72	260	2.5	1.8	9.0	2.4
Germany	79	460	3.2	2.0	11.9	3.1
Berlin U.[g]		22	2.2	0.7	5.6	
Italy	75	90	1.4	1.2	5.0	1.7
Netherlands	(70)	55	2.6	1.8	6.1	2.7
Leiden U.[h]		10	1.3	0.8	5.3	
United States[i]	67	160	1.1	0.8	5.5	1.2
California U.[j]		3.6	0.5		3	
Columbia U.[k]		10	0.8	0.7	10	
Harvard U.[l]		14	2.0	1.1	5	

[a]The number of papers is the product of the "best guess" of the GNP of physical papers (Table E.1) and the percent thereof due to affiliates of our set of higher schools in the country in question (Table A.8, column 1, to which we have added half the contributions by unidentified contributors).

[b]The numbers of physicists are taken from Table A.1 for countries as a whole and Table A.4 for individual institutions.

[c]Total expenditure from Table I; likewise laboratory expenditure for countries as a whole. The indices for individual institutes are based on their regular annual budgets only (Table C.2); i.e., they ignore income from extraordinary appropriations, grants, etc. The indices for the United States are per 1.5×10^4 marks expended.

[d]Cavendish Lab., *History* (1910), pp. 281-334.

[e]Manchester, U., *Phys. Labs.* (1906), pp. 63–124.

[f]Toronto, U., *1827-1906,* pp. 230–55, gives twelve papers in the seven years 1897–1903.

[g]Rubens in Lenz, *Gesch. U. Berlin, 3* (1910), 292; U. Berlin, *Chronik* (1900–04).

Physicists at the T.H.'s were evidently no less productive of pure physics papers than their colleagues at the universities. Of the 76 academic contributors to the *Annalen der Physik* in 1900 21 percent were affiliated with the T.H.'s; of our 145 post-holding German academic physicists 25 percent were affiliated with T.H.'s. The small difference is more than accounted for by the absence of research students at the T.H.'s.

2. Input: Resources Devoted to Research

The productivities computed in Table E.3 are based on gross inputs. Even were there no discrepancies between the figures for the resources expended upon academic physicists in the several countries and the rates of production of their research papers, much interest would attach to the proportion of resources devoted to research, to the cost, in time and money, of producing a physics paper. And although these discrepancies are remarkably small for the three leading European nations it is nonetheless clear that one man-year, or 10,000 marks, spent upon physics does not buy the same number of papers everywhere. In view of the great difficulties in testing hypotheses about the relation of input of funds to output of papers, one might be tempted simply to assume a rough proportionality between input and output, and to take productivity (papers per man or mark) as a direct measure of the proportions of available resources applied to research in the several countries. We do, however, have not only our own intuitive judgments but also a limited amount of data bearing directly upon this question. We have therefore permitted ourselves to guess the proportions of each category of support which the academic physicists devoted to research in Germany, France, Britain, and the United States. Our results appear in Table E.4, the least reliable of all our tables.

a. Time.

The amount of time a physicist devoted to research depended strongly upon his nationality and his rank.[6] The chairholder in Germany each

[6]In Italy professors were supposed to be appointed on the basis of "scientific attainments and didactic aptitude." Oldrini, U.S. Office of Ed., *Report, 1* (1901-02), 777. For Britain see Thomson, *Recollections* (1937), p. 61, and in Brit. Ass. Adv.

---

[h]Leiden, U., Phys. Lab., *Communications* (1898-1902).

[i]We have *not* used here the factor 3/2 by which the number of affiliates and the funding of physics at our set of 21 American universities was increased to obtain the national total in our previous tables, since this factor, appearing in both numerator and denominator, cancels in productivity computations. However, we have multiplied all U.S. expenditures by a factor of 0.67 to take account of the lower research purchasing power in the U.S. In 1962, this factor ranged between 0.53 (relative to the Netherlands) and 0.67 (relative to France). Nat. Res. Coun., Phys. Survey Comm., *Phys. in Perspective* (1972), pp. 541-42.

[j]Birge, "Phys. Dept., Berkeley," *1* (1966), v (30).

[k]*Columbia Quarterly, 1* (1898-99), 217; *2* (1899-1900), 420-21; *3* (1900-01), 421; *4* (1901-02), 442-43; *5* (1902-03), 490. It must be said, however, that the majority of the publications were slight or slightly popular.

[l]*Contributions from the Jefferson Physical Laboratory, 1* (1903)—*6* (1908) suggest 14 papers in 1900.

TABLE E.4

Resources Devoted to Research (1000's of Marks)

	British Empire		France		Germany		United States[a]	
	Percent	Amount	Percent	Amount	Percent	Amount	Percent	Amount
A. Time devoted to research	38	65	46	67	57	133	27[b]	53
Chair holders	20	8	20	5	30	11	20	7
Junior faculty	30	10	40	10	40	12	25	8
Assistants & Privatdozenten	40	23	40	20	50	38	30	25
Other affiliated researchers	60	24	80	32	80	72	25	13
B. Personal income	3	29	4	24	3	23	3	29
Salary, fees, private income	3	29	3	19	3	23	3	29
Prizes	—	—	35	5	—	—	—	—
C. Laboratory expenditure	50	158	38	110	57	220	40	175
Institute budget	45	121	35	88	50	130	35[c]	135
Special intramural grants	60	12	40	12	65	65	50	10
Research grants	100	5	100	2	100	5	100	10
Private philanthropy	100	20	100	8	100	20	100	20
D. Laboratory construction	35	270	35	85	40	135	20[d]	120
Total of B+C+D	22	460	19	220	25	380	16	325

[a]The factor 3/2 has *not* been introduced; the figures are for our set of 21 American institutions.

[b]Cooke, *Acad. Efficiency* (1910), Table 7, yields 26 to 50 percent as the range of the proportion of teaching salary expense chargeable to research in four leading American universities (Columbia, Harvard, Princeton, Wisconsin), and 29 percent for his entire sample of eight colleges and universities, in 1909. The breakdown by ranks for his entire sample is: professors 23 percent, asst. professors 33 percent, instructors 26 percent, assistants 32 percent. In making our guess for 1900 we have sought to allow for the rapid increase in emphasis upon research in American universities. Thus only one among the ten graduate students in physics at Harvard in 1894/5 was engaged in research, but just five years later six of the ten graduate students were so engaged. J. Trowbridge in Harvard U., *President's Report* (1894–95), pp. 207–08; (1899–1900), p. 253.

[c]Cooke, *loc. cit.*, yields 36 to 62 percent for his four universities, averaging 40 percent for his entire sample, in 1909. This breaks down as: 60–80 percent, and 66 percent, for wages for mechanics, etc.; 48–65 percent, and 48 percent, for equipment and supplies; 16–34 percent, and 20 percent, for maintenance.

[d]Cooke, *loc. cit.*, yields 38 to 75 percent for the proportion of the value of "physical equipment" chargeable to research in his four universities, 44 percent for his entire sample, in 1909; 16 to 33 percent for the proportion of the value of "land, building and fixture" devoted to physics which is chargeable to research in his four universities, 17 percent for his entire sample.

semester gave a lecture course of four or five hours a week with elaborate experimental demonstrations. Preparing these lectures, even with the help of an assistant, could consume much time which otherwise would have been devoted to research.[7] Yet on the whole the German professor probably did not spend much more time teaching than did the French, who might take advantage of his light load of three hours of lecturing to accumulate additional posts. The teaching duties of the British chairholder seem to have been higher than both, and those of the American highest of all.[8]

Aside from the examination of his own students (and of others in Britain), the chairholder had to sit on a variety of government committees regulating admission to schools, scholarships, certification of teachers, engineers, and medical students. In Britain and Germany, this service brought in significant additions to the professor's income, included

---

Sci., *Adv. of Sci.* (1931), p. 6; Tilden, *William Ramsay* (1918), pp. 69, 83; Grant, *Bragg* (1952), p. 16; Eve, *Rutherford* (1939), p. 52, 59, 137. For a debate on teaching vs. research see Tilden, *Nature, 64* (1901), 585; Ramsay, *Nature, 64* (1901), 388-91; *Nature, 61* (1900), 395-96. For the very rapid growth of research interests in the U.S. around 1900 see Kevles, *Phys. in America* (1964), pp. 145-49, 181-87, 200-02, 265, 288-89, and Reingold, "American Indifference" (1971). Cf. the "Statement" by W.F. Magie regarding appointments to be made of junior faculty at Princeton: "Every opportunity, which can be afforded in the physical laboratory, will be given to the occupants of these positions to engage in research and to influence and direct such advanced students as may be working in the Department." (Magie to H. Crew, 26 May 1906, Crew Papers, AIP.) For Germany see Paulsen, *Autobiography* (1938), chs. 10-12. Paschen wrote Kayser, 29 Nov. 1910, that a candidate for a post as *assistant* must have completed some independent research. (Staatsbibl. Preuss. Kulturbesitz, Berlin-Dahlem). In France, research was considered "la fonction la plus essentielle des Facultés des Sciences." Paris, U., Conseil acad., *Rapports* (1902-03), p. 105; and again, France, Min. instr. publ., *Enquêtes,* Nr. 101 (1911), pp. 22-23; see Paul, *French Hist. Studies,* 7 (1972), 423-26.

[7]This was considered the most demanding of all duties. Küchler, "Phys. Labs. Germany" (1906), p. 193. Paschen feared that the preparation of such a course of lectures at Tübingen would take all his time. Paschen to Kayser, 18 Feb. 1901. (Staatsbibl. Preuss. Kulturbesitz, Berlin-Dahlem.) Cf. Section B, supra.

[8]Liard, *Enseignement supérieur, 2* (1894), 512; Cotton, *Les Curie* (1963), pp. 63-64; Langevin, *Revue du mois, 2* (1906), 21. An extreme case was that of Lodge, who complained that for a while at Liverpool, "the work was getting beyond even my energy. I sometimes lectured five hours a day; even the standing up all that time was tiring." Lodge, *Past Years* (1931), p. 159. See Thomson, *Recollections* (1937), p. 129; Eve, *Rutherford* (1939), pp. 64, 69; Kay, *Nat. Phil., 1* (1963), 136-38. Cooke, *Acad. Efficiency* (1910), Table 4. Böttger, *Amer. Hochschulwesen* (1906), pp. 51-52, gives an opposite view of the relative burdens of the German and American academics. The Italian professor lectured 6 to 10 hours per week; the Italian Privatdocent 3 to 6. See, e.g., Pisa, U., *Annuario* (1899-1900); Rome, U., *Annuario* (1899-1900), pp. 134-35; Turin, U., *Annuario* (1898-99).

as far as possible in Table B.1; it also occasioned a share of the complaints that the professor had no time for research.[9] We judge that among chairholders the Germans devoted more of their time during the semester to research, including therein the supervision of research students, than did their colleagues in other countries.

In none of these countries, so far as we can judge, were the academic vacations specifically devoted to research, either by the chairholder or by anyone else in his institute.[10] And when research work was done, the level of activity fell far below that in term-time. Both the terms and the hours of work in term-time were longer in Germany and the United States than in Britain and France.[11]

The low incomes of junior faculty and assistants encouraged moonlighting rather than research in the free hours.[12] We suspect that among junior faculty the Germans and French gave most time to research, primarily because they had the lightest teaching duties, but also because they knew that their chances of advancement depended heavily upon their publications. This applies a fortiori to the German Privatdozenten. In general the Germans seem to have been able to resist temptation to do outside work, both because of independent means and their appreciation of the cost to their careers. The assistant, if not also a Privatdozent, could devote to research roughly the half of each day not taken up by his official duties. Whether he did so depended largely on the expectations in his academic environment.[13] We suppose that other researchers affiliated with the institutes—students, overwhelmingly—worked longer hours in Germany and the United States than elsewhere, since that was generally expected.

[9] E.g., Liebig to A.W. Hofmann, 14 Nov. 1863: "What wouldn't I give to be able to shake off the mind-deadening examinations of the medical and pharmacy students here in Munich, but in Berlin the examinations are annihilating." Heinig, *Forschen u. Wirken, 1* (1960), 34; Appell, *Revue de Paris, 17* (1910), 120.

[10] Paulsen, *German Univ.* (1906), p. 301; Kayser, "Erinnerungen" (1936), passim.

[11] "A laboratory in this country [U.S.] in which nobody ever began work before 10 a.m. or worked later than 6 in the evening would serve as a terrible example of sloth and indolence." Bumstead to N.R. Campbell referring to the Cavendish, quoted in Cavendish Lab., *History* (1910), pp. 225-26. The University of Edinburgh physical laboratory was open daily from 10 a.m. to 3 p.m. *Calendar* (1900), p. 65. Cf. the comments on Röntgen by Friedrich, interview 15 May 1963, pp. 11-13 (AHQP).

[12] Langevin, Rutherford, and Einstein all tried tutoring in order to scrape by. Langevin, *Langevin* (1971), pp. 45-46; Eve, *Rutherford* (1939), pp. 34, 39; cf. Eulenburg, *Nachwuchs* (1908), pp. 146–49. Cf. notes 43 and 44 to section B, supra.

[13] Moving a salary increase for assistants, twenty-two Austrian parliamentarians asserted that "these years, in which Wilhelm Ostwald places the most important contributions of a scientist, are in many cases a period of deprivation and hunger. The as-

b. Personal Income.

Physicists of 1900 often pictured their researches as paid for chiefly from personal income—a gross distortion, but not without some semblance of truth. Paschen, celebrating his collaboration with Carl Runge in the years 1895 to 1901, recalled: "The diffraction gratings used in this work were Runge's property. Whatever else was required but not available in the institute we purchased out of our own pockets."[14] The grating had indeed been purchased by Runge, but as his accounts show, his research expenses in 1893 and 1894 had been entirely covered by a grant from the Berlin Academy, and doubtless the goodly surplus and subsequent grants covered the expenses in the following years as well. Yet from about 1905 to 1915 Paschen himself spent more than 3,500 marks from his salary and fees to purchase equipment for his researches. It is likely that a significant fraction of his colleagues approached his level of expenditure, and a few far exceeded it.[15] Such scattered evidence is the basis for our entries in Table E.4. Of particular interest are prizes for completed research, which we have treated as personal income, but of which a substantial fraction doubtless found its way into research.[16]

---

sistant is thus compelled to seek additional employment. He finds it, however, only in the rarest cases, and when he does he sacrifices his extremely limited free time and endangers thereby his future; for his advancement depends wholly and solely upon his scientific contributions." Austria, Reichsrat, Abgeordnetenhaus, *Sten. Prot.,* 21. Sess. (1912), Beilage 1287.

[14]Paschen, "Runge," *Naturw., 15* (1927), 231. In two years Kayser and Runge paid out 314 marks to five firms from their grant. Runge to Kayser, 22 Dec. 1894 (Staatsbibl. preuss. Kulturbesitz, Berlin-Dahlem).

[15]As part of Paschen's compensation for refusing a call to Göttingen the Württemberg Kultusministerium agreed in 1915 to buy from him for the Tübingen physical institute research instruments valued at 3,500 marks which Paschen "had purchased at his own expense because of the insufficiency of the budget funds" (U. Archiv, Tübingen, 117/904). Around 1910 Otto Lehmann had made a similar deal with the Baden Kultusministerium, resulting in the purchase from him for the Karlsruhe physical institute of research instruments in the amount of 20,000 marks. Lehmann estimated that he had paid three times that amount for them. *Phys. Inst. Karlsruhe* (1911), p. 74. This comes to 3,000 marks per year out of pocket. K. Birkeland used his private income to support his two assistants circa 1910 and to pay other research expenses amounting in some years to as much as 13,000 marks. Oslo, U., personal communication (21 March 1973). See also the remarks on private laboratories in Section A.

[16]For example, the Nobel prizes of Rayleigh and the Curies; supra Section C, note 76. Blondlot gave 60 percent of his 50,000 franc Leconte prize for scholarships in physics and mathematics. Nancy U., *1854-1904,* p. 134.

c. Laboratory Expenditure.

The most important source of support for physics research was the budgets of the physics institutes, in particular those entries designated for the construction and purchase of apparatus and supplies. Access to this resource, which in European laboratories lay under the virtually exclusive control of the professor-director, was the sine qua non of physics research for our affiliated physicists. Very few institute budgets contained funds specifically designated for research purposes. In France and Germany, certainly, the nominal purpose of the budgets, just as of the institutes themselves and their cadre of assistants, was to provide for practical instruction.[17]

Much of the apparatus as well as important features of the electrical installation of the German institute, although ostensibly in the interest of striking and up-to-date lecture demonstrations, was in fact requested or purchased by the institute director principally with an eye to research.[18] In France "through bookkeeping transfers, 'dishonest' ones if you will, our laboratory directors diverted to research sums intended for teaching."[19] The bureaucracy knew, but tolerated, the situation; official reports freely admitted that advanced students and professors did research in all the Sorbonne laboratories, even the P.C.N. ones, "although they are laboratories for instruction."[20] In Britain, on the other hand,

[17]Some English and American laboratories had funds specifically labelled for research; in France Lippmann's laboratory was itself specifically for research, and in that sense so was its budget.

[18]Thus Paschen (21 April 1901, U. Archiv, Tübingen, 117/904), taking over the Tübingen institute, requested i.a. a large Rowland grating (1,500 marks) *not* because it was absolutely indispensable for the continuation of his researches, but because he would not be able to *demonstrate* the Zeeman effect, "one of the most interesting phenomena of modern physics," without one. Indeed Paschen explicitly averred that everything he requested was "solely" for instructional purposes, not for his own researches.

[19]Perrin (1936), in Lot, ed., *Perrin* (1936), pp. 162-67. See also Perrin, *Science et espérance* (1948), p. 91. Pierre Curie's research in the 1890's was supported, "thanks to the kindness of his superiors," from the teaching laboratory fund. M. Curie, *Pierre Curie* (1923), p. 28. Langevin's large thesis credits only Bouty, "who gave me the use of the precious resources of the physics teaching laboratory at the Sorbonne." Langevin, *Recherches* (1902), p. 5. Cf. Fabry, *Annales de l'Univ. de Paris, 3* (1928), 551. Of 11,000 francs spent on equipment and gas at the physics laboratory of the École normale in 1893 more than two-thirds were ascribed to the costs of research. Violle to the Director of the École normale, 22 Dec. 1893, as reported by H. Abraham, 9 Jan. 1901. (AN/61 AJ 117).

[20]Paris, U., Conseil acad., *Rapports* (1902-03), p. 110; see also (1910-11), p. 159; for bureaucrats turning a blind eye to the diversion of teaching funds see Gilpin, *France Sci. State* (1968), pp. 96-97.

with no direct governmental oversight, the laboratory director seems to have had less need for dissimulation when appropriating funds to research.[21]

Even where a specific budget appropriation for research existed, one cannot infer from it the proportion of laboratory expenditure devoted to the purpose. Crucial pieces of apparatus often came on loan from other laboratories, or in trade for goods or services.[22] Furthermore one must take into account the extensive use of teaching apparatus for research purposes, and national variations in the practice;[23] and one must prorate appropriately *all* entries in the institute budget, from the wages of the mechanic to the heating of the building. Cooke's close examination of the total research expenditures in the physics departments of eight leading U.S. colleges and universities in 1909[24] allows us to proceed from qualitative intuitions about the relative place of research in the several countries to quantitative estimates of the proportion of their institute budgets effectively applied to it (Table E.4): Germany 50 percent, France 35, Britain 45, United States 35. These fractions roughly equal our estimates of the proportion of time which the assistants spent upon research in the several countries; the efforts of this class of affiliates, so closely associated with the institutes in both their instructional and research aspects, are likely to be a good index of the balance of activities in the institute as a whole.

Finally we add the funds specifically intended for the support of research, the grants and philanthropy of outside agencies. In Table E.4 we

[21] Newall in Cavendish Lab., *History* (1910), p. 109; J.J. Thomson, quoted in Rayleigh, *Thomson* (1942), p. 18.

[22] E.g., Rutherford and Soddy used radiothorium placed at their disposal by Knöfler of Berlin. Rutherford, *Collected Papers, 1* (1962), 435. Becquerel used liquid air given by d'Arsonval. *CR, 133* (1901), 199-202, 977-80. Runge loaned his diffraction grating to the Göttingen institute. Voigt in Göttingen, *Phys. Inst.* (1906), pp. 41-42. The Curies loaned radium to various parties in France and abroad. M. Curie, *Pierre Curie* (1928), p. 102.

[23] Rowland wrote D.C. Gilman, 5 Nov. 1875: "You were right when you said I would find no lack of scientific spirit here [in Berlin], and the apparatus shows it. In America we have apparatus for illustration, in England and France they have apparatus for illustration and experiment, but in Germany, they have *only* apparatus for experimental investigation." Quoted by J.D. Miller, *Isis, 63* (1972), 10.

[24] Cooke, *Acad. Efficiency* (1910), Table 7 et passim. J.J. Thomson, *Recollections* (1937), pp. 125-26, gave 11,400 marks as an upper limit for the "extra expense" incurred by the Cavendish Laboratory in consequence of the research pursued in 1912. This figure when compared with the matériel budget for 1910 (Table C.2) suggests that research took around half the budgeted funds. Cf. Thomson in Brit. Ass. Adv. Sci., *Adv. of Sci.* (1931), p. 12.

TABLE E.5

Productivity of Research by Academic Physicists, circa 1900[a]

	Papers per man-year of research by:		Papers per 10^4 marks (1.5×10^4 in U.S.) of:	
	Post-holding physicists	All affiliated physicists	Laboratory expenditure on research	Total expenditure on research
British Empire	7.1	4.5	18	6.3
France	7.4	3.9	24	12
Germany	7.5	3.5	21	12
United States	4.0	3.0	14	7.5

[a]Here as in Table E.3 we have allowed a factor of 0.67 for the lower research purchasing power of the mark in the U.S.: thus the U.S. figures are papers per 15,000 marks. The numerator of these indices, annual number of publications, is drawn from Table E.3; the denominators from Table E.4.

take over directly the figures given in Table C.1 and compute the total "laboratory expenditure" devoted to research.

d. New Plant.

In row D of Table E.4 we assign to research a portion of the cost of new buildings and their initial outfitting with equipment and apparatus. These estimates are based once again upon Cooke's study of American institutions and our intuitions about European institutions relative to them. We have also been guided in part by crude estimates of the fraction of the floor space plausibly assignable to research in the European institutes; with space as with equipment, research and teaching funds overlapped.

3. Productivities of Physics Research

Having made an independent estimate of the time and money devoted to physics research in the institutions of higher education of the four leading nations, we conclude by comparing those estimates with the measures of output offered above. We compute a new set of productivity indices (Table E.5) using as the unit of input not the total resources supplied physicists, but the fraction applied to research (Table E.4).[25]

[25]The results obtained in Table E.5 are a little disappointing. Our failure to achieve substantial equality in the values of the indices for the several countries may be attributed either to inaccuracy of our estimates of relative research effort (Table E.4), or to inapplicability of our assumption of a direct proportionality between input of

The output of papers per unit of funding, whether gross (Table E.3) or net (Table E.5), is far more uniform from nation to nation when taken relative to laboratory expenditure than when measured by the total expenditure on academic physics; in the short run that category of funding was decisive for the quantity of research produced. Doubtless the other categories—buildings, salaries—play an important, possibly dominant, role in the long run.

The uniformity of our productivity indices—i.e., the circumstance that each research paper, whether German, French, or British, represented about the same amount of time and money expended—reinforces the proposition suggested at the outset by the uniformity of our investment indices (relative to population and national product): by 1900 scientific life in physics had assumed, at least in Northern Europe, an international character. We are naturally unable to say anything on the basis of this study about the relative contribution of each paper to the progress of physics, or about the existence and significance of national gifts and styles.

---

time and/or money and output of papers, independent of nationality. We incline to the first alternative. In particular, the considerable gap remaining between the American and European values we construe as a failure on our part to appreciate how very small a fraction of the available resources were devoted to research in the United States in 1900.

BIBLIOGRAPHY

The bibliography gives in full the references cited in the text. In the case of articles in collaborative histories, *Festschriften,* etc., the title of the article is given in full under the author's name, with an abbreviated reference to the author and title of the collaborative work. A full bibliographical entry for the work will be found under its own "author," often an institution. Cross references are provided under the name of the institution to all works dealing particularly with one of our higher schools.

Aachen. Technische Hochschule. *Rheinisch-Westfälische Technische Hochschule Aachen, 1870–1970.* Ed. H.M. Klingenberg. 2 vols. Stuttgart: Bek, 1970.

Ångström, Knut. "Fysiska institutionen." Uppsala. Univ. *Uppsala universitet, 2* (1897), 148–55.

Aberdeen. University. "Abstract of Accounts" and "Statistical Report." [Annual, as follows:] Great Britain. Parliament. House of Commons. *Sessional Papers,* 1902, *81,* 235–82; 1903, *53,* 453–500; 1904, 77, 69–120; 1905, *61,* 195–246; 1906, *92,* 481–532; 1907, *65,* 547–98; 1908, *86,* 823–72; 1909, *69,* 535–86; 1910, *72,* 657–708; 1911, *60,* 1–54.

——. ——. *Calendar.* [Annual.]

——. ——. *Handbook to City and University: Quatercentenary Celebrations, September 1906.* Aberdeen: [Univ. ?], 1906.

Aberystwyth. University. See: Great Britain, Board of Education, *Report;* Great Britain, Committee on the University of Wales, *Report.*

[Abraham, H.] *Henri Abraham. Commémoration du centenaire de sa naissance à l'École Normale Supérieure le 7 décembre 1968.* [Paris: École Normale ?], 1969.

Académie Royale des Sciences. Brussels. *Règlements.* Brussels: Hayez, 1896.

Adelaide. University. *Calendar.* [Annual.]

Adressbuch der lebenden Physiker, Mathematiker und Astronomen des In- und Auslandes und der technischen Hilfskräfte. 2nd ed. Leipzig: Barth, 1909.

Allen, Frank, and William Ambler. "Test of the liquid air plant at Cornell University." *Physical Review, 15* (1902), 181–87.

Alsace-Lorraine. Landesausschuss. "Uebersicht der ordentlichen Einnahmen und Ausgaben bei der Kaiser-Wilhelms-Universität Strassburg." *Landeshaushalts-Etat von Elsass-Lothringen.* [Annual.]

American Academy of Arts and Sciences. *The Rumford Fund of the American Academy of Arts and Sciences.* Boston: The Academy, 1905.

Amsterdam. Universiteit. *Gedenkboek van het Athenaeum en de universiteit van Amsterdam, 1632–1932.* Amsterdam: Stadsdrukkerij, 1932.

Annuaire des sociétés savantes, littéraires et artistiques de Paris. Ed. Réunion des secrétaires généraux. Paris: Inst. International de Bibliographie Scientifique, 1898.

Appell, Paul. "La Faculté des sciences de l'Université de Paris (1895–1910)." *La Revue de Paris, 17,* No. 6 (1910), 98–120.

———. "L'Enseignement supérieur des sciences." *Revue générale des sciences, 15* (1904), 287–99.

———. *Souvenirs d'un Alsacien.* Paris: Payot, 1923.

Armagnat, H. "Les Appareils de mesures électriques à l'exposition." *Journal de physique, 10* (1901), 165–95.

Auerbach, Felix. *Ernst Abbe, sein Leben, sein Wirken, seine Persönlichkeit; nach den Quellen* . . . Leipzig: Akademische Verlagsges., 1918.

Austria. Bundesministerium für Unterricht. *100 Jahre Unterrichts-Ministerium, 1848–1948. Festschrift.* Vienna: Österreichischer Bundesverlag, 1948.

———. Ministerium für Kultus und Unterricht. *Neubauten für Zwecke des naturwissenschaftlichen, medizinischen, technischen und landwirtschaftlichen Unterrichtes an den Hochschulen in Wien, 1894–1913:* Im Auftrag des k.k. Ministers für Kultus und Unterricht dargestellt [by Ludwig Cwiklinski] und der in Wien vom 21. bis 26. September 1913 tagenden 85. Versammlung Deutscher Naturforscher und Ärzte gewidmet. Vienna: Hof- und Staatsdruckerei, 1913.

———. Reichsrat. Haus der Abgeordneten. "Interpellation . . . betreffend endliche Erbauung eines physikalischen Institutsgebäudes an der k.k. Universität in Wien." *Stenographische Protokolle,* 18. Session, Anhang, vol. 7 (82. Sitzung, 4 June 1908), pp. 9791–92 (Interpellation 2998).

———. ———. ———. "Antrag . . . betreffend die Regelung der Gehaltsverhältnisse der Assistenten." *Stenographische Protokolle,* 21. Session (1912), Beilage 1287.

Ayrton, W.E. "The Education of the American Electrical Engineer." *Mosely Educational Commission to the United States of America, Oct.–Dec. 1903: Reports.* London: Cooperative Printing Society, 1904. Pp. 26–39.

Badash, Lawrence. "An Elster and Geitel Failure: Magnetic Deflection of Beta Rays." *Centaurus, 11* (1966), 236–40.

———. "Nagaoka to Rutherford, 22 February 1911." *Physics Today, 20* (1967), 55-60.

Badash, Lawrence, ed. *Rutherford and Boltwood: Letters on Radioactivity.* New Haven: Yale University Press, 1969.

Baden. Stände-Versammlung. Zweite Kammer. "Denkschrift über die künftige bauliche Entwickelung der badischen Hochschulen." *Verhandlungen,* Heft 502 (1913), 31-41.

———. ———. ———. "Spezial-Budget des Ministeriums der Justiz, des Kultus und Unterrichts für die Jahre 1900 und 1901." *Verhandlungen,* Landtag 1899-1900, Beilagenheft 3 [Drucksachensammlung Heft 455].

———. ———. ———. "Staatsvoranschlag 1910 und 1911." *Verhandlungen,* Landtag 1909-1910. [Drucksachensammlung Heft 490].

Bailey, Kenneth C. *A History of Trinity College Dublin 1892-1945.* Dublin: University Press, 1947.

Baker, Marcus. "Endowments for Research in the United States." Carnegie Institution of Washington. *Report of Executive Committee. November 11, 1902* [Washington D.C., 1902], pp. 247-69.

Bangor. University College of North Wales. *Calendar.* [Annual.]

———. ———. See: Great Britain, Board of Education, *Report;* Great Britain, Committee on the University of Wales, *Report.*

Barff, H.E. *A Short Historical Account of the University of Sydney . . . in Connection with the Jubilee Celebrations, 1852-1902.* Sydney: Angus & Robertson, 1902.

Barus, Carl. "The Progress of Physics in the Nineteenth Century." *Congress of Arts and Science . . . St. Louis 1904.* Ed. H.J. Rogers. New York: Houghton, Mifflin & Co., 1906. Vol. 4, 29-65.

Basel (canton). *Staats-Rechnung.* [Annual.]

———. Universität. *Personal-Verzeichnis.* [Annual.]

———. ———. See: Hagenbach-Bischoff, ed.; Lebhardt, A; Thommen, R.

Bastiani, R. "I nuovi istituti scientifici dell'Ateneo bolognese." *Giornale del genio civile* [= *Annali dei lavori pubblici*], *45* (1907), 667-92.

Baumgarten, Fritz. *Freiburg im Breisgau. Der Albert-Ludwigs-Universität Freiburg zur Feier ihres 450 jährigen Bestehens.* Die deutschen Hochschulen. Band 1. Berlin: Wedekind & Co., 1907.

Bavaria. Landtag. Kammer der Abgeordneten. "Etat des k. Staatsministeriums des Innern für Kirchen- und Schulangelegenheiten." *Verhandlungen,* 33. Landtag, Beilagen-Bd. 4, Budget Nr. 26 (for 25. Finanzperiode, 1900 and 1901), Beilagen-Bd. 10, Budget Nr. 26 (for 26. Finanzperiode, 1902 and 1903), Beilagen-Bd. 15, Budget Nr. 26 (for 27. Finanzperiode, 1904 and 1905); 34. Landtag, Beilagen-Bd. 4, Budget Nr.

26 (for 28. Finanzperiode, 1906 and 1907); 35. Landtag, Beilagen-Bd. 5, Budget Nr. 28 (for 1908 and 1909), Beilagen-Bd. 12, Budget Nr. 28 (for 1910 and 1911).

Becherer, Gerhard. "Die Geschichte der Entwicklung des Physikalischen Instituts der Universität Rostock." *Wissenschaftliche Zeitschrift der Universität Rostock, Math.-Naturwiss. Reihe, 16* (1967), 825-37.

Beckman, Anna, and Per Ohlin. *Forskning och undervisning i fysik vid Uppsala Universitet under fem århundraden: En kortfattad historik.* (*Acta universitatis upsaliensis.* Skrifter rörande Uppsala universitet. C. Organisation och historia, no. 8.) Uppsala, 1965.

Beier, Adolf. *Die höheren Schulen in Preussen . . . und ihre Lehrer.* 3rd ed. Halle: Waisenhaus, 1909.

Belfast. Queens College. *Report of the President.* [Annual.]

——. ——. See: Moody, T.W.

Bellivier, A. *Henri Poincaré ou la vocation souveraine.* Paris: Gallimard, 1956.

Bellot, H. Hale. *University College, London, 1826-1926.* London: University of London Press, 1929.

Ben-David, Joseph. "Scientific Productivity and Academic Organisation." *American Sociological Review, 25* (1960), 828-43.

——. *The Scientist's Role in Society.* Englewood Cliffs, N.J.: Prentice-Hall, 1971.

Ben-David, Joseph, and Abraham Zloczower. "Universities and Academic Systems in Modern Societies." *European Journal of Sociology, 3* (1962), 45-84.

Benham, W.B. "The London 'Times' on the Scientific Situation." *Nature, 53* (1895), 73-74.

Berget, Alphonse. "Le Nouveau laboratoire des recherches physiques de la Sorbonne." *La Nature, 26* (1898), 225-27.

Berlin. Landwirtschaftliche Hochschule. See: Börnstein, R.

——. Technische Hochschule. *Die Königliche Technische Hochschule zu Berlin.* Berlin: Mückenberger, 1903.

——. ——. *Die Technische Hochschule zu Berlin, 1799-1924. Festschrift.* Berlin: Georg Stilke, 1925.

——. ——. See: Borchardt, R.; *ZBBV, 21.*

——. Universität. *Amtliches Verzeichnis des Personals und der Studierenden.* [Semiannual.]

——. ——. *Forschen und Wirken: Festschrift zur 150 Jahr-Feier der Humboldt-Universität zu Berlin.* 2 vols. Berlin: VEB Deutscher Verlag der Wissenschaften, 1960.

———. ———. See: Guttstadt, A.; Kleinwächter, F.; Lenz, M.; Rubens, H.

Bern (canton). *Staats-Rechnung.* [Annual.]

———. Direktion des Unterrichtswesens. *Verwaltungsbericht.* [Annual.]

———. Universität. *Behörden, Lehrer und Studierende.* [Annual.]

———. ———. *Die naturwissenschaftlichen und medicinischen Institute der Universität Bern, 1896.* Biel: Schüler, [1896].

———. ———. See: Feller, R.

Besançon. Université. *L'Université de Besançon: ses origines, son organisation actuelle.* Besançon: Dodivers, 1900.

Biquard, Pierre. *Langevin.* Savants du monde entier, v. 38. Paris: Seghers, 1969.

Birge, Raymond Thayer. "History of the Physics Department, [University of California, Berkeley]." 4 vols. mimeographed. Berkeley, 1966.

Birkenhead, Earl of. *The Prof in Two Worlds: The Official Life of Professor F.A. Lindemann, Viscount Cherwell.* London: Collins, 1961.

Birmingham. University. *Calendar.* [Annual.]

———. ———. See: Great Britain, Board of Education, *Report.*

Bishop, Morris. *A History of Cornell.* Ithaca: Cornell University Press, 1962.

Bluntschli, Friedrich, and G. Lasius. "Der neue Physikbau für das eidgenössische Polytechnikum zu Zürich." *Schweizerische Bauzeitung, 10* (1887), 9, 22–24, and plates facing pp. 6, 12, 20.

Boeck, Carl, ed. *Deutsche Technisch-Wissenschaftliche Forschungsstätten.* Teil II: *Die Technisch-Wissenschaftlichen Forschungsanstalten.* Berlin: VDI-Verlag, 1931.

Börnstein, R. "Das neuerbaute physikalische Institut der Landwirtschaftlichen Hochschule in Berlin." *Physikalische Zeitschrift, 12* (1911), 551–58.

———. "Das Physikalische Institut der Landwirtschaftlichen Hochschule in Berlin." Deutsche Physikalische Ges., *Verhandl., 13* (1911), 206–08.

Böttger, Wilhelm Carl. *Amerikanisches Hochschulwesen: Eindrücke und Betrachtungen.* Leipzig: Engelmann, 1906.

du Bois, Henri. "Die Einrichtung physikalischer Privatlaboratorien." *Elster-Geitel Festschrift* (1915), pp. 245–50.

———. "Halbring-Elektromagnet." *Zeitschrift für Instrumentenkunde, 19* (1899), 357–64.

Bologna. Università. *Stabilimenti scientifici della R. università di Bologna in rapporto col piano regolatore della città secondo il progetto del rettore G. Capellini.* Bologna: Succ. Monti, 1888.

———. ———. *L'université de Bologne autrefois et aujourd'hui. Par les soins d'un comité de professeurs de la même université*. Trans. A. de Carli. Bologna: Zanichelli, [1921].

——. ——. See: Bastiani, F.

Bonn. Universität. *Chronik*. [Annual.]

——. ——. *Geschichte der Rheinischen Friedrich-Wilhelms-Universität zu Bonn am Rhein*. Ed. A. Dyroff. Vol 2: *Institute und Seminare, 1818–1933*. Bonn: F. Cohen, 1933.

——. ——. See: Kayser, H.; Konen, H.

Borchardt, R. "Die Hundertjahrfeier der Technischen Hochschule in Berlin." *Physikalische Zeitschrift, 1* (1899), 70–72.

Borel, Emil. "Paul Villard." *Comptes rendus, 199* (1934), 1467.

Bose, Margrete. "Das Physikalische Institut der Universität La Plata." *Physikalische Zeitschrift, 12* (1911), 1230–43.

du Bourguet, L. "Notice sur la Faculté des Sciences de Marseille." Marseille. Université. Faculté des Sciences. *Annales, 10* (1900), i–xxii.

Brandstaetter, F. "Die Geschichte der physikalischen Institute seit 1815." Vienna. Technische Hochschule. *150 Jahre, 2* (1965), 159–178.

Brauer, Ludolf et al., eds. *Forschungsinstitute: Ihre Geschichte, Organisation und Ziele*. 2 vols. Hamburg: P. Hartung, 1930.

Braunschweig. Technische Hochschule. *Programm*. [Annual.]

——. ——. See: Schneider, W.

Breslau. Technische Hochschule. *Technische Hochschule in Breslau. Festschrift zur Eröffnung 1. Oktober 1910*. Breslau: F. Hirt, [1910].

——. Universität. *Chronik*. [Annual.]

——. ——. *Festschrift zur Feier des hundertjährigen Bestehens der Universität Breslau*. Teil 2: *Geschichte der Fächer, Institute und Ämter der Universität Breslau, 1811–1911*. Breslau: F. Hirt, 1911.

——. ——. *Personal-Verzeichnis*. [Semiannual.]

——. ——. See: Lummer, O.; Meyer, O.E.; *ZBBV, 23*.

Brillouin, Marcel. "Les débuts de la Société française de physique." *Le livre du cinquantenaire de la Société française de physique*. Paris: Revue d'Optique Théorique et Instrumentale, 1925. Pp. 3–18.

Bristol. University College. *Calendar*. [Annual.]

——. ——. *Meeting of the Governors and Report of the Council*. [Annual.]

——. ——. See: Great Britain, Board of Education, *Report*.

Broca, André. "L'oeuvre d'Henri Becquerel." *Revue générale des sciences, 19* (1908), 802–13.

Broda, Engelbert. *Ludwig Boltzmann: Mensch, Physiker, Philosoph*. Vienna: Franz Deuticke, 1955.

Brown, Addison. "The Need of Endowment for Scientific Research and Publication." U.S. Office of Education. *Report* (1890–91), *2*, 1060–66. Reprinted: *Nature, 51* (1894), 164–67, 186–89.

Brüche, Ernst. "Aus der Vergangenheit der Physikalischen Gesellschaft. III." *Physikalische Blätter, 17* (1961), 27–33.

Brünn. Technische Hochschule. *Festschrift der k.k. Technischen Hochschule in Brünn zur Feier ihres fünfzigjährigen Bestehens und der Vollendung des Erweiterungsbaues im Oktober 1899.* Brünn: Rohrer, 1899.

Brunhes, Bernard. "La Physique dans les universités italiennes." *RIE, 41* (1901), 400–05.

Burchardt, Lothar. "Wissenschaftspolitik und Reformdiskussion im Wilhelminischen Deutschland." *Konstanzer Blätter für Hochschulfragen, 8* (1970), 71–84.

Burdeau, ?. "Le Budget de l'instruction publique." *RIE, 23* (1892), 170–71.

Burke, John Butler. "The Physical Laboratory at the Muséum d'Histoire Naturelle." *Nature, 71* (1904), 177–79.

Busch, Alexander. *Die Geschichte des Privatdozenten.* Göttinger Abhandlungen zur Soziologie, Band 5. Stuttgart: Ferdinand Enke, 1959.

Cabannes, Jean. "Hommage à Henri Abraham, Eugène Bloch, Georges Bruhat . . . Morts en déportation." *Annales de l'Université de Paris, 22* (1952), 368–82.

Cagliari. Università. *Annuario.*

Cajori, Florian. *A History of Physics.* 2nd ed. New York: Dover, 1962.

California. University. *President's Report.* [Annual.]

———. ———. *Register.* [Annual.]

Callot, Jean-Pierre. *Histoire de l'École Polytechnique.* Paris: Les Presses Modernes, 1959.

Cambridge. University. *The Historical Register of the University of Cambridge* [to 1910]. Ed. J.R. Tanner. Cambridge: Cambridge University Press, 1917. *Supplement, 1911–1920.* Cambridge: Cambridge University Press, 1922.

———. ———. See: Clark, J.W.; Howard, H.F.; *Nature, 75, 78;* Previté-Orton, C.W.; Roth, L.; Rothblatt, S.; Rouse Ball, W.W.; Rumsey, C.A.; Sviedrys, R.; Winstanley, D.A.

———. ———. Cavendish Laboratory. *A History of the Cavendish Laboratory, 1871–1910.* London: Longmans, Green, 1910.

Cambridge Economic History of Europe. Vol. 6: *The Industrial Revolutions and After: Incomes, Population and Technological Change.* Ed.

H.J. Habakkuk and M. Postan. Cambridge: Cambridge University Press, 1966.

Canada. Royal Commission on Industrial Training and Technical Education. *Report.* 3 vols. Ottawa: Parmelee, 1913.

Cant, Ronald Gordon. *The University of St. Andrews: A Short History.* Edinburgh: Oliver and Boyd, 1946.

Cardiff. University College of South Wales and Monmouthshire. *Calendar.* [Annual.]

——. ——. See: Great Britain, Board of Education, *Report;* Great Britain, Committee on the University of Wales, *Report.*

Cardwell, Donald Stephen Lowell. *The Organization of Science in England.* Melbourne: Heinemann, 1957.

Carhart, Henry S. "The Imperial Physico-Technical Institution at Charlottenburg." *Science, 12* (1900), 697–708. Also in: Smithsonian Institution. *Annual Report* (1900), pp. 403–15.

Carman, Albert P. "The Design of a Physical Laboratory." *Brickbuilder, 20* (1911), 257–60. (Reprinted in *Some Contributions from the Laboratory of Physics of the University of Illinois, Urbana, 1912–1914,* pp. 10–16.)

Carnegie Endowment for International Peace. *A Manual of the Public Benefactions of Andrew Carnegie.* Washington, D.C.: CEIP, 1919.

Carnegie Foundation for the Advancement of Teaching. *Bulletin,* No. 1: *Papers Relating to the Admission of State Institutes to the System of Retiring Allowances of the Carnegie Foundation.* New York: CFAT, 1907.

——. *Bulletin,* No. 2. See: *The Financial Status of the Professor in America and in Germany.* New York: CFAT, 1907.

——. *Bulletin,* No. 5. See: Cooke, M.L.

Carnegie Institution of Washington. [*"Confidential"*] *Report of* [*the*] *Executive Committee* [*to the Board of Trustees*], *November 11, 1902.* [Washington, D.C.: Carnegie Inst., 1902].

——. *Yearbook.* Washington, D.C.: Carnegie Inst., 1902.

Casimir, H.B.G., and J.H. Van Vleck. *Cherwell-Simon Memorial Lectures 1961 and 1962.* Edinburgh: Oliver and Boyd, 1962.

Catania. Università. *Storia della università di Catania dalle origini ai giorni nostri.* Catania: Zuccarello & Izzi, 1934.

——. ——. See: Mandalari, M.

Cattell, J. McKeen. *American Men of Science.* 1st ed. New York: Science Press, 1906.

——. "The Scientific Men of the World." *The Scientific Monthly, 23* (1926), 468–71.

——. "A Statistical Study of American Men of Science" and "A Further Statistical Study of American Men of Science." *American Men of Science.* 2nd ed. New York: Science Press, 1910. Pp. 537–96.

Caullery, Maurice. "L'Evolution de notre enseignement supérieur scientifique." *Revue du mois, 4* (1907), 513–35.

——. *La Science française depuis le XVIIIe siècle.* Paris: Colin, 1933.

——. *Universities and Scientific Life in the United States.* Cambridge Mass.: Harvard Univ. Press, 1922. Translation of the French edition, Paris, 1917.

Cavendish Laboratory. See: Cambridge. University. Cavendish Laboratory.

Chapman, Allan. "Notes on the cost of physics research and teaching at Oxford University over the years 1900–1910." Personal communication, March 1973.

Chapman, Arthur W. *The Story of a Modern University: A History of the University of Sheffield.* London: Oxford University Press, 1955.

Charlottenburg. Physikalisch-Technische Reichsanstalt. *Die bisherige Tätigkeit der Physikalisch-Technischen Reichsanstalt: Aus der dem Reichstage am 19. Februar 1904 überreichten Denkschrift, mit einem Verzeichnis der Veröffentlichungen aus den Jahren 1901–1903.* Ed. F. Kohlrausch. Braunschweig: Vieweg, 1904.

——. ——. *Forschung und Prüfung: 50 Jahre Physikalisch-Technische Reichsanstalt.* Ed. J. Stark. Leipzig: S. Hirzel, 1937.

——. ——. *Wissenschaftliche Abhandlungen.* Berlin: J. Springer, 1894–.

——. ——. "Die Thätigkeit der Physikalisch-Technischen Reichsanstalt." *Zeitschrift für Instrumentenkunde,* 1891–.

——. ——. See: Carhart, H.S.; Pfetsch, F.; Voigt, W.; Warburg, E.

Charpentier, J.L. "Les Médecins et l'idéal scientifique à la scène." *Revue du mois, 6* (1908), 457–80.

Chicago. University. *President's Report.* [Annual.]

——. ——. *Register.* [Annual.]

——. ——. See: Goodspeed, T.W.; Ryan, W.C.

——. ——. Graduate School. *Register.* [Annual.]

Cincinnati. University. See: Dwight, C.H.

Clark, John Willis, ed. *Endowments of the University of Cambridge.* Cambridge: Cambridge University Press, 1904.

Clark, Terry N., and P.P. Clark. "Le patron et son cercle: Clef de l'université française." *Revue française de sociologie, 12* (1971), 19–39.

Clark University. *President's Report.* [Annual.]

——. *Register and Official Announcements.* [Annual.]

——. See: Ryan, W.C.; Webster, A.G.

Claude, Georges. *Ma vie et mes inventions.* Paris: Plon, 1957.

Coben, Stanley. "The Scientific Establishment and the Transmission of Quantum Mechanics to the United States, 1919–1932." *American Historical Review, 76* (1971), 442–466.

Cochrane, Rexmond. *Measures for Progress: A History of the National Bureau of Standards.* Washington, D.C.: N.B.S., 1966.

Cohen, John Michael. *The Life of Ludwig Mond.* London: Methuen, 1956.

Cohen, Robert S., and Raymond J. Seeger, eds. *Ernst Mach: Physicist and Philosopher.* Boston Studies in the Philosophy of Science, vol. 6. Dordrecht: Reidel, 1970.

Coletti, Francesco. "La classificazione statistica per età dei professori ordinari e straordinari delle regie università e dei regii istituti superiori." Istituto lombardo di scienze e lettere. *Rendiconti, 51* (1918), 513–22.

Colson, C. "La Préparation aux écoles techniques supérieures." *Revue générale des sciences, 15* (1904), 299–303.

Columbia University. *Catalogue.* [Annual.]

____. *President's Report.* [Annual.]

____. *Treasurer's Report.* [Annual.]

____. See: Hallock, Wm.

Comité International des Poids et Mesures. *Procès-Verbaux.* Paris: Gauthier-Villars, 1877–.

Commonwealth Universities Yearbook (original title *The Yearbook of the Universities of the Empire).* London: Universities Bureau, 1914–.

Congrès International de Physique, réuni à Paris en 1900. *Rapports.* Ed. C.E. Guillaume and L. Poincaré. 4 vols. Paris: Gauthier-Villars, 1900–01.

Conrad, Johannes. *The German Universities for the Last Fifty Years.* Trans. John Hutchison. Glasgow: David Bryce and Son, 1885.

____. "Einige Ergebnisse der deutschen Universitätsstatistik." *Jahrbücher für Nationalökonomie und Statistik, 32* (1906), 433–92.

____. *Das Universitätsstudium in Deutschland während der letzten 50 Jahre: Statistische Untersuchung* . . . . Jena: G. Fischer, 1884.

Cooke, Morris Llewellyn. *Academic and Industrial Efficiency.* Carnegie Foundation for the Advancement of Teaching, Bulletin No. 5. New York: CFAT, 1910.

Cope, Alexis. *History of the Ohio State University.* Vol. 1: *1870–1910.* Columbus: The Ohio State Univ. Press, 1920.

Copenhagen. *Aarbog for Københavns universitet, kommunitetet og den polytekniske laereanstalt.* Copenhagen: Universitetsbogtrykkeriet. [Triennial.]

____. Polytekniske laereanstalt. *Den Polytekniske laereanstalt: Samlinger, laboratorier m.m.* Supplement til laereanstaltens program. Copenhagen: J.M. Schultz, 1910.

____. ____. See: Lundbye, J.

____. Universitet. See: Copenhagen; *RIE, 26.*

Corbino, O.M. "Il contributo italiano ai progressi della Elettrologia nell' ultimo cinquantennio." Società italiana per il progresso delle scienze. *Atti, 5* (1911), 275–306.

Cork. Queens College. *Report of the President.* [Annual.]

Cornell University. *President's Report.* [Annual.]

____. *Register.* [Annual.]

____. *Treasurer's Report.* [Annual.]

____. See: Allen, F.; Bishop, M.; Hewett, W.T.; Howe, H.E.

Cory, Ralph. "Fifty Years at the Royal Institution." *Nature, 166* (1950), 1049–53.

Cotton, Eugénie. *Aimé Cotton.* Savants du monde entier, v. 36. Paris: Seghers, 1967.

____. *Les Curie.* Savants du monde entier, v. 14. Paris: Seghers, 1963.

Cram, Ralph Adams. "Princeton Architecture." *The American Architect, 96,* (1909), 21–30.

Crew, Henry. "Diary, 1895." Crew Papers, Niels Bohr Library, American Institute of Physics, New York.

Crozier, Michel. *The Bureaucratic Phenomenon.* Chicago: University of Chicago Press, 1964.

Curie, Eve. *Madame Curie.* Garden City, New York: Doubleday, Doran, 1937.

Curie, Marie. *Pierre Curie.* 1923. New York: Dover, 1963.

Curie, Pierre. *Oeuvres.* Paris: Gauthier-Villars, 1908.

Curti, Merle, and Roderick Nash. *Philanthropy in the Shaping of American Higher Education.* New Brunswick, New Jersey: Rutgers University Press, 1965.

Curzon, George Nathaniel. *Principles and Methods of University Reform.* Oxford: The Clarendon Press, 1909.

Czernowitz. Universität. *Die k.k. Franz-Josephs-Universität in Czernowitz im ersten Vierteljahrhundert ihres Bestandes. Festschrift.* Czernowitz: Bukowinaer Vereinsdr., 1900.

____. ____. *Übersicht der akademischen Behörden, Professoren, Privatdozenten, Lehrer, Beamten, etc.* [Annual.]

Dalhousie. University. *Calendar.* [Annual.]

____. ____. See: Johnstone, J.H.L.

Damm, Paul Friedrich. *Die Technischen Hochschulen in Preussen: Eine Darstellung ihrer Geschichte und Organisation.* Berlin: Mittler, 1899.

——. *Die technischen Hochschulen Preussens: Ihre Entwicklung und gegenwärtige Verfassung . . . nach amtlichen Quellen.* Berlin: Mittler, 1909.

Danne, Jacques. *Notice sur le Laboratoire d'Essais des Substances radioactives à Gif (Seine-et-Oise).* 1912. As translated in *Physikalische Zeitschrift, 13* (1912), 565–74.

Danzig. Technische Hochschule. *Technische Hochschule in Danzig: Festschrift zur Eröffnung, 6. Oktober 1904.* Danzig: T.H., 1904.

Darboux, Gaston. *Éloges académiques et discours.* Paris: Hermann, 1912.

——. "Rapport présenté au conseil supérieur de l'instruction publique sur un projet de décret relatif au doctorat ès sciences." *RIE, 35* (1898), 263–67.

Darlington, Thomas. *Education in Russia.* Great Britain, Board of Education, Special Reports on Educational Subjects, vol. 23. London: HMSO, 1909. Also in Great Britain. Parliament. House of Commons. *Sessional Papers,* 1909 (Cd. 4812), *18,* 495–1069.

Darmstadt. Technische Hochschule. *Die neuen Gebäude der grossherzoglichen technischen Hochschule zu Darmstadt: Festschrift zur feierlichen Einweihung der Neubauten am 28. Oktober 1895.* Darmstadt: [T.H.], 1895.

——. ——. *Programm.* [Annual.]

——. ——. *Die Technische Hochschule Darmstadt, 1836 bis 1936: Ein Bild ihres Werdens und Wirkens.* [Darmstadt] : W. Schlink, [1936].

Dartmouth College. See: Nichols, E.F.

Davies, William, and W. Lewis Jones. *The University of Wales and its Constituent Colleges.* London: F.E. Robinson & Co., 1905.

[DBF.] *Dictionnaire de biographie française.* Ed. J. Balteau et al. Paris: Letouzly, 1933–.

Deane, Phyllis, and W.A. Cole. *British Economic Growth, 1688–1959.* 2nd ed. Cambridge: Cambridge Univ. Press, 1967.

Delaunay, Henri. *Annuaire international des sociétés savantes pour 1903.* Paris: A. Lahure, 1904.

Delft. Technische Hoogeschool. *Gedenkschrift van de koninklijke akademie en van de polytechnische school 1842–1905. Samengesteld ter gelegenheid van de oprichting der technische hoogeschool.* Delft: Waltman, 1906.

De-Metts, Georgii Georgievich. "Fizicheskie instituty i masterskiia fizicheskikh priborov za granitseiu." *Inzhener, 23* (1899), 462–66, 498–

506. [The continuations of this article in vol. 24 (1900) have not been seen by us.]

_____. *Physical Laboratory of the Polytechnic Institute Emperor Alexander II at Kiev: Construction and Interior Installations.* [In Russian.] Kiev, 1903. [Not seen.]

Denmark. Danmarks Statistik. *Statistisk Årbog: 1972: Statistical Yearbook.* Vol. 76. Copenhagen: Danmarks Statistik, 1972.

Dennison, David M. "Physics and the Department of Physics since 1900." *Research-Definitions and Reflections: Essays on the Occasion of the University of Michigan Sesquicentennial.* Ann Arbor: [U. of Michigan], 1967. Pp. 120–36.

Des Coudres, Theodor. "Das theoretisch-physikalische Institut." Leipzig. Univ. *Festschrift, 4,* Pt. 2 (1909), 60–69.

Deslandres, H. "Actions mutuelles des électrodes et des rayons cathodiques dans les gaz raréfiés." *Comptes rendus, 124* (1897), 678–81.

_____. "Allocution." Société française de physique. *Bulletin, 37* (1909), 2*–4*.

Douglas, A.V. *The Life of Arthur Stanley Eddington.* London: Nelson, 1956.

Dresden. Technische Hochschule. *Die Neubauten der Königl. Sächsischen Technischen Hochschule zu Dresden.* Berlin: A.W. Schade, 1905.

[DSB.] *Dictionary of Scientific Biography.* Ed. C.C. Gillespie. New York: Scribner, 1970–.

Dublin. University. *Calendar.* [Annual.]

_____. _____. See: Bailey, K.

Dubos, René J. *Louis Pasteur: Freelance of Science.* Boston: Little, Brown, 1950.

Dundee. University College. *Calendar.* [Annual.]

Dupree, A. Hunter. *Science in the Federal Government: A History of Policies and Activities to 1940.* Cambridge, Mass.: Harvard Univ. Press, 1957.

Dupuy, Gaston. "Sur la vie et les travaux de André Debierne." *Bulletin de la société chimique de France, 17* (1950), 1023–26.

Durm, Josef. "Das physikalische und physiologische Institut der Universität Freiburg." *Zentralblatt der Bauverwaltung, 13* (1893), 93–95.

Duruy, Victor. *Notes et souvenirs, 1811–1894.* 2 vols. Paris: Hachette, 1902.

Dushman, Saul. *Scientific Foundations of Vacuum Technique.* New York: Wiley, 1949.

Dwight, C. Harrison. "The First Seventy-five Years of the Physics Department (1883–1958)" [of the University of Cincinnati]. Cincinnati, 1969. Niels Bohr Library, American Institute of Physics, New York.

Edinburgh. University. "Abstract of Accounts" and "Statistical Report." [Annual, as follows:] Great Britain. Parliament. House of Commons. *Sessional Papers,* 1902, *81,* 283–310; 1903, *53,* 501–26; 1904, *77,* 121–46; 1905, *61,* 247–72; 1906, *92,* 533–60; 1907, *65,* 599–624; 1908, *86,* 873–98; 1909, *69,* 587–612; 1910, *72,* 709–34; 1911, *60,* 55–90.

——. ——. *Calendar.* [Annual.]

——. ——. *History of the University of Edinburgh 1883–1933.* Ed. A. Logan Turner. Edinburgh: Oliver and Boyd, 1933.

——. ——. See: *Nature.*

[Elster-Geitel.] *Arbeiten aus den Gebieten der Physik, Mathematik, Chemie. Festschrift Julius Elster und Hans Geitel . . . .* Braunschweig: F. Vieweg & Son, 1915.

Enriques, Federigo. "L'università italiana. Critica degli ordinamenti in vigore." *Scientia, 3* (1908), 133–47.

Erikson, Henry A., with John Zeleny. "The University of Minnesota Department of Physics." No date. Niels Bohr Library, American Institute of Physics, New York.

Erlangen. Universität. *Uebersicht des Personal-Standes.* [Annual.]

——. ——. See: Kolde, T.; Wiedemann, E.

Erman, Wilhelm, and Ewald Horn. *Bibliographie der deutschen Universitäten: Systematisch geordnetes Verzeichnis der bis Ende 1899 gedruckten Bücher und Aufsätze über das deutsche Universitätswesen.* 3 vols. Leipzig: B.G. Teubner, 1904.

Etard, M.A. "Notice sur la vie et les travaux de E.A. Demarçay." *Bulletin de la société chimique de Paris, 31* (1904), i–viii.

Eulenburg, Franz. *Der "Akademische Nachwuchs": Eine Untersuchung über die Lage und die Aufgaben der Extraordinarien und Privatdozenten.* Leipzig: B.G. Teubner, 1908.

——. *Die Frequenz der deutschen Universitäten von ihrer Gründung bis zur Gegenwart.* Leipzig: B.G. Teubner, 1904. (Akademie der Wissenschaften. Leipzig. Philologisch-Historische Klasse. *Abhandlungen, 24,* No. 2 (1904).)

——. "Das Alter der deutschen Universitätsprofessoren." *Jahrbücher für Nationalökonomie und Statistik, 25* (1903), 65–80.

Eve, A.S. "Some Scientific Centres. VIII. The MacDonald Physics Building, McGill University, Montreal." *Nature, 74* (1906), 272–75.

———. *Rutherford. Being the Life and Letters of the Rt. Hon. Lord Rutherford, O.M.* New York: MacMillan, 1939.

Fabry, Ch. "La Physique à Paris: enseignement et laboratoires." *Annales de l'Université de Paris, 3,* No. 6 (1928), 551–75.

Falkus, M.E. "Russia's National Income, 1913: A Revaluation." *Economica, 35* (1968), 52–73.

Feller, Richard. *Die Universität Bern, 1834–1934.* Bern: Paul Haupt, 1935.

Ferber, Christian von. *Die Entwicklung des Lehrkörpers in deutschen Universitäten und Hochschulen, 1865–1954.* Untersuchungen zur Lage der deutschen Hochschullehrer. Vol. 3. Göttingen: Vandenhoek und Ruprecht, 1956.

Ferraris, Carlo Francesco. *Cinque anni di rettorato nella R. università di Padova, 1891–1892 al 1895–1896: Ricordi in occasione del settimo centenario.* Rome: Stabilimento poligrafico per l'amministrazione dalla guerra, 1922.

———. "Statistica dei consorzi universitari italiani." R. istituto veneto di scienze, lettere ed arti, *Atti, 61,* No. 2 (1901–02), 305–14.

———. "Laureati e diplomati nelle università e negli istituti superiori italiani nel quinquenio scolastico dal 1904–1905 al 1908–1909." *Ibid., 70,* No. 2 (1910), 205–28.

———. "Gli inscritti nelle università e negli istituti superiori italiani nel decennio scolastico dal 1893–1894 al 1902–1903." *La riforma sociale, 13* (1903), 877–903, 965–77.

———. "Gli inscritti nelle università e negli istituti superiori italiani nel quattordicennio dal 1893–1894 al 1906–1907." *Ibid., 17* (1907), 733–40.

———. *Statistiche delle Università e degli Istituti superiori.* Italy. Istituto centrale di statistica. *Annali di Statistica,* Ser. V, vol. 6. Rome: Tipogr. Nazionale, 1913.

Feuer, Lewis S. "The Social Roots of Einstein's Theory of Relativity." *Annals of Science, 27* (1971), 277–98, 313–44.

Financial Status of the Professor in America and in Germany. Carnegie Foundation for the Advancement of Teaching, *Bulletin,* No. 2. New York: CFAT, 1907.

Flach, Johannes. *Der deutsche Professor in der Gegenwart.* Leipzig: A. Unblad, 1886.

Forcrand, R. de. "Les Instituts scientifiques et les nouvelles universités." *Revue générale des sciences, 8* (1897), 613–17.

Forman, Paul. *The Environment and Practice of Atomic Physics in Weimar Germany.* Diss. Univ. Calif., Berkeley, 1967. Ann Arbor: University Microfilms, 1968.

——. "Financial Support and Political Alignment of Physicists in Weimar Germany." *Minerva, 12* (1974), 39–66.

Forman, Paul, and A. Hermann. "Sommerfeld, Arnold." *DSB* (in press).

Fournier d'Albe, E.E. *The Life of Sir William Crookes.* London: T. Fisher Unwin, 1923.

Foville, A. de. "La Caisse des recherches scientifiques." *Revue scientifique, 49* (1911), 385–88.

Fowler, Alfred. "The Equipment of the Spectroscopic Laboratory of the Imperial College of Science and Technology." Physical Society of London, *Proceedings, 24* (1912), 168–71.

Fowler, J.T. *Durham University. Earlier Foundations and Present Colleges.* London: Robinson and Co., 1904.

France. Assemblée Nationale. Chambre des Députés. *Débats.*

——. Documentation Française. "La France devant les problèmes de la recherche." *Notes et études documentaires,* Nos. 2552 and 2580 (1959).

——. Institut National de la Statistique et des Études Économiques. *Annuaire Statistique,* v. 58. Paris: Imprimerie Nationale, 1951.

——. *Journal Officiel de la République Française.*

——. Ministère de l'Éducation Nationale. Personal communication from Maurice Bayen, Dépt. de l'histoire de l'éducation, Ministère de l'éducation nationale, Paris, 13 April 1973.

——. Ministère de l'Instruction Publique. *Catalogue des thèses (1899–1900).* Paris: Librairie Hachette, 1900.

——. ——. "Rapports des conseils des universités pour l'année scolaire." *Enquêtes et documents relatifs à l'enseignement supérieur.* [Annual. As follows: No. 101 (1909–10), No. 104 (1910–11), No. 106 (1911–12), No. 108 (1912–13).]

——. ——. *Recueil des lois et règlements sur l'enseignement supérieur, 4* (1884–89), *5* (1889–98), *6* (1898–1909), 7 (1909–14).

——. ——. *La science française.* Exposition universelle et internationale de San Francisco. Paris: Min. I. P., 1915.

——. ——. *Statistique de l'enseignement supérieur.* Vol. 3: *1878–1888.* Paris: Imprimerie nationale, 1889. Vol. 4: *1889–1899.* Paris: Imprimerie nationale, 1900.

——. ——. Caisse des recherches scientifiques. *Rapports scientifiques sur les travaux entrepris en 1904.* Melun: Imprimerie administrative, 1905.

Frankfurt a.M. Physikalischer Verein. *Der Neubau des Physikalischen Vereins und seine Eröffnungsfeier am 11. Januar 1908.* Frankfurt a.M.: Naumann, 1908.

____. ____. See: Wachsmuth, R.

Freiburg i. Br. Universität. *Verzeichnis der Behörden, Lehrer, Anstalten, Beamten.* [Annual.]

____. ____. See: Baumgarten, F.; Durm, Josef.

French, John C. *A History of the University Founded by Johns Hopkins.* Baltimore: Johns Hopkins Press, 1946.

Fribourg (canton). Direction de l'instruction publique. *Compte rendu.* [Annual.]

____. Université. *Autorités, professeurs et étudiants.* [Annual.]

____. ____. *Bericht.* [Annual.]

Gaede, Hannah. *Wolfgang Gaede.* Karlsruhe: Braun, 1954.

Gaede, Wolfgang. "Die Molekularluftpumpe." *Naturwissenschaften, 1* (1913), 11-14.

____. "Hochvakuumpumpe nach Gaede." *Zeitschrift für Instrumentenkunde, 27* (1907), 163-65.

Galton, Douglas, et al. "On the Establishment of a National Physical Laboratory. Report of the Committee." British Association for the Advancement of Science. *Report* (1896), 82-85.

Galway. Queen's College. *Calendar.* [Annual.]

____. ____. *The Report of the President.* [Annual.]

Gauja, Pierre. *Les fondations de l'Académie des sciences, 1881-1915.* Hedaye, France: Imprimerie de l'observatoire d'Abbadia, 1917.

Gebhardt, Willy. "Die Geschichte der Physikalischen Institute der Universität Halle." Halle. Univ. *Wissensch. Zeitschr., Math. Nat. Reihe, 10* (1961), 851-60.

General Education Board. *The General Education Board: An Account of its Activities, 1902-1914.* New York: G.E.B., 1915.

Geneva (canton). *Budget.* [Annual.]

____. Université. *Actes du jubilé de 1909.* Geneva: Librairie Georg, 1910.

____. ____. *Historique des facultés, 1896-1914.* Geneva: Georg, 1914.

____. ____. *Liste des autorités, professeurs, étudiants et auditeurs.* [Annual.]

Genoa. Università. *L'Università di Genova.* Genoa: Stab. ital. arti grafiche, [1923].

Gerbod, Paul. *La condition universitaire en France au XIXe siècle.* Paris: Presses Universitaires, 1965.

Giessen. Universität. *Personenbestand.* [Semiannual.]

——. ——. See: Lorey, W.; Wien, W.

Gilpin, Robert George. *France in the Age of the Scientific State.* Princeton: Princeton Univ. Press, 1968.

Glaisher, J.W.L. "The Mathematical Tripos." London Mathematical Society. *Proceedings, 18* (1886), 4-38.

Glasgow. University. "Abstract of Accounts" and "Statistical Report." [Annual, as follows:] Great Britain. Parliament. House of Commons. *Sessional Papers,* 1902, *81,* 311-50; 1903, *53,* 527-68; 1904, *77,* 147-88; 1905, *61,* 273-314; 1906, *92,* 561-604; 1907, *65,* 625-68; 1908, *86,* 889-942; 1909, *69,* 613-58; 1910, *72,* 735-80; 1911, *60,* 91-156.

——. ——. *Calendar.* [Annual.]

——. ——. See: Gray, A.; Mackie, J.D.

Glasser, Otto. *Dr. W.C. Röntgen.* 2nd ed. Springfield, Ill.: Charles C. Thomas, 1958.

——. *Wilhelm Conrad Röntgen and the Early History of Roentgen Rays.* Springfield, Ill.: Charles C. Thomas, 1934.

——. *Wilhelm Conrad Röntgen und die Geschichte der Röntgenstrahlen.* 2nd ed. Berlin: Springer, 1959.

Glazebrook, R.T. "The Aims of the National Physical Laboratory of Great Britain." Smithsonian Institution. *Annual Report* (1901), pp. 341-57.

Glum, Friedrich. "Die Kaiser–Wilhelm–Gesellschaft zur Förderung der Wissenschaft: ihre Forschungsaufgaben, ihre Institute und ihre Organisation." *Forschungsinstitute* . . . . Ed. L. Brauer, et al., *1* (1930), 359-73.

——. *Zwischen Wissenschaft, Wirtschaft und Politik.* Bonn: Bouvier, 1964.

Goblet d'Alveilla, Comte E.-F.-A. *L'Université de Bruxelles pendant son troisième quart de siècle.* Brussels: Weissenbruch, 1909.

Göttingen. Universität. *Amtliches Verzeichnis des Personals und der Studierenden.* [Semiannual.]

——. ——. *Chronik.* [Annual.]

——. ——. *Die Physikalischen Institute der Universität Göttingen: Festschrift im Anschlusse an die Einweihung der Neubauten am 9. Dezember 1905.* Leipzig, Berlin: G.B. Teubner, 1906.

——. ——. See: Klein, F.; Riecke, E.

Goldberg, Stanley. "History of Physics at Harvard University 1907-1912." Cambridge, Mass., 1962. Niels Bohr Library, American Institute of Physics, New York.

Gollob, Hedwig. *Geschichte der Technischen Hochschule in Wien; nach neugefundenem Aktenmaterial bearbeitet.* Wien: Gerold and Co., 1964.

Goodspeed, Thomas W. *A History of the University of Chicago: The First Quarter Century.* Chicago: Univ. Chicago Press, 1916.

Goodwin, H.M. "Physics at M.I.T. A History of the Department from 1865-1933." *The Technology Review, 35* (1933), 287-91, 312-13.

Gosselet, J. "L'Enseignement des sciences appliquées dans les universités." *RIE, 37* (1899), 97-106.

Grant, Kerr. *The Life and Work of Sir William Bragg.* Brisbane: Univ. Queensland Press, 1952.

Gray, Andrew. "Famous Scientific Workshops. I: Lord Kelvin's Laboratory in the University of Glasgow." *Nature, 55* (1897), 486-92.

Gray, James. *The University of Minnesota, 1851-1951.* Minneapolis: University of Minnesota Press, 1951.

Graz. Universität. *Verzeichnis der Akademischen Behörden, Lehrer und Beamten.* [Annual.]

____ . ____ . See: Rezori, W.E. von.

Great Britain. Board of Education. *Report for the Year 1901, on the Museums, Colleges and Institutions, under the Administration of the Board of Education.* London: HMSO [Cd. 1266], 1902.

____ . ____ . *Reports from those Universities and University Colleges in Great Britain which Participate in the Parliamentary Grant for the University Colleges.* [Annual. All have command numbers and most, though published separately, also have volume numbers in Great Britain, Parliament, House of Commons, *Sessional Papers (SP),* as follows:] 1899-1900 (London, 1901), Cd. 845; 1900-01 (1902), Cd. 1510 (*SP*, 1903, *20*, 573-768); 1901-02 (1904), Cd. 1888 (*SP*, 1904, *19*, 465-680); 1902-03 (1904), Cd. 2366 (*SP*, 1905, *25*, 103-324); 1903-04 (1905), Cd. 2789 (*SP*, 1906, *28*, 191-436); 1904-05 (1906), Cd. 3409 (*SP*, 1907, *22*, 161-496); 1905-06 (1907), Cd. 3885 (*SP*, 1908, *26*, 121-511); 1906-07 (1908), Cd. 4440 (*SP*, 1908, *26*, 511-938); 1907-08 (1909), Cd. 4879 (*SP*, 1909, *19*, 149-662); 1908-09 (1910), Cd. 5246 (*SP*, 1910, *24*, 121-731); 1909-10 (1911), Cd. 5872; 1910-11 (2 vols. 1912), Cd. 6245, 6246; 1911-12 (2 vols. 1913), Cd. 7008, 7009; 1912-13 (2 vols. 1914), Cd. 7614, 7615; 1913-14 (2 vols. 1915), Cd. 8137, 8138.

____ . Committee on the University of Wales. "Minutes of Evidence." Great Britain. Parliament. House of Commons. *Sessional Papers,* 1909, *19* (Cd. 4572), 699-838,

____ . ____ . "Report." *Sessional Papers,* 1909, *19* (Cd. 4571). 663-98.

____ . National Physical Laboratory. *Report.* Teddington: Parrott and Ashfield, 1905-.

———. ———. See: Glazebrook, R.T.; Great Britain, Parliament, House of Commons; Langdon-Davies, J.

———. Parliament. House of Commons. *General Alphabetical Index to the Bills, Reports, Estimates, Accounts, and Papers, Printed by Order of the House of Commons, and to the Papers Presented by Command, 1852–1899.* London: HMSO, 1909.

———. ———. ———. *General Index to the Bills, Reports and Papers Printed by Order of the House of Commons and to the Reports and Papers Presented by Command 1900 to 1948–1949.* London: HMSO, 1960.

———. ———. ———. *Sessional Papers.* As cited below, in the following form: year of session (Command number of doc.), *volume in which doc. appears,* pagination within vol. (Pagination within doc.)

———. ———. ———. "Return from the Universities of Oxford and Cambridge separately, stating, so far as possible for each year, from 1870–1875, inclusive, number, names, and description of professors and mode of appointment of each professor, emoluments of each professor, distinguishing the amounts accruing from fees, salary and other sources." *Sessional Papers,* 1876 (Cd. 327), *59.*

———. ———. ———. "Report of the Committee Appointed by the Treasury to Consider the Desirability of Establishing a National Physical Laboratory [and] Minutes of Evidence Taken before the Committee . . . ." *Sessional Papers,* 1898 (Cd. 8976, 8977), *45,* 337–463 (1–10, i–iv, 1–113).

———. ———. ———. "Preliminary Report of the Departmental Committee on the Royal College of Science . . . ." *Sessional Papers,* 1905 (Cd. 2610), *61,* 423–33 (1–10).

———. ———. ———. "Final Report of the Departmental Committee on the Royal College of Science . . . Minutes of Evidence . . . ." *Sessional Papers,* 1906 (Cd. 2872, 2956), *31,* 391–665 (i–vi, 1–33, i–iv, 1–231).

the Last Fifty Years, towards Establishing, Endowing, and Maintaining the Various Scientific Societies in (1) England, (2) Scotland, and (3) Ireland." *Sessional Papers* 1906, *65* (H.C. 358), 515–17.

———. ———. ———. "National Physical Laboratory: Accounts of Receipts and Expenditure . . . 1900 to 1906." *Sessional Papers,* 1907, *68* (H.C. 116), 361–83.

———. ———. ———. "First Annual Report of the Governing Body of the Imperial College of Science and Technology . . . ." *Sessional Papers,* 1909 (Cd. 4602), *19,* 123–49 (i–ii, 1–24).

———. ———. ———. "Committee on the University of Wales and its Constituent Colleges: Minutes of Evidence." *Sessional Papers,* 1909 (Cd. 4572), *19,* 699–838 (i–ii, 1–138).

———. ———. ———. "Accounts of Receipts and Expenditures of the Universities and Colleges, Ireland. . . ." *Sessional Papers,* 1911 (H.L. 347), *59,* 385-402 (1-18).

Gregory, Winifred, comp. *List of the Serial Publications of Foreign Governments, 1815-1931.* New York: H.W. Wilson Co., 1932.

Greifswald. Universität. *Amtliches Verzeichnis des Personals.* [Annual.]

———. ———. *Chronik.* [Annual.]

———. ———. *Festschrift zur 500-Jahrfeier der Universität Greifswald 17. 10. 1956.* 2 vols. Greifswald: Universität, 1956.

———. ———. *Verzeichnis der Vorlesungen.* [Annual.]

———. ———. See: Schallreuter, W.; *ZBBV, 11.*

Grenoble. Université. *Annuaire.*

———. ———. *Université de Grenoble: établissements universitaires, personnel, organisation de l'enseignement, statistique des étudiants.* Grenoble, 1900.

———. ———. *Université de Grenoble 1339-1939.* Grenoble: Allier, 1939.

Griewank, Karl. *Staat und Wissenschaft im Deutschen Reich: Zur Geschichte und Organisation der Wissenschaftspflege in Deutschland.* Freiburg i. Br.: Herder, 1927.

Groningen. Universiteit. *Academia groningana, 1614-1914. Gedenkboek ter gelegenheid van het derde eeufeest der Universiteit te Groningen.* Groningen: Noordhoff, 1914.

———. ———. *Universitas groningana, 1614-1964. Gedenkboek ter gelegenheid van het 350-jarig bestaan.* 2 vols. Groningen: Wolters, 1964-66.

———. ———. See: Haga, H.

Guerlac, Henry. "Science and French National Strength." *Modern France.* Ed. Edward M. Earle. Princeton: Princeton Univ. Press, 1951. Pp. 81-105.

Guillaume, Ch. "E. Mascart." *La Nature, 36* (1908), 238-40.

Guttstadt, Albert, ed. *Die naturwissenschaftlichen und medicinischen Staatsanstalten Berlins: Festschrift für die 59. Versammlung deutscher Naturforscher und Aerzte.* Berlin: August Hirschwald, 1886.

Guye, Charles-Eugène. "L'école polytechnique fédérale de Zurich." *Revue générale des sciences, 8* (1897), 102-09.

de Haas-Lorentz, Gertrude L., ed. *H.A. Lorentz. Impressions of his Life and Work.* Amsterdam: North-Holland, 1957.

Haga, Hermann. "Het natuurkundig laboratorium." Groningen. Universiteit. *Academia Groningana 1614-1914* (1914), pp. 485-93.

Hagenbach-Bischoff, Eduard. "Die Entwicklung der naturwissenschaftlichen Anstalten Basels 1817-1892." Schweizerische Naturforschende Gesellschaft. *Verhandlungen, 75* (1892), 1-36.

Haguenin, E. "L'Université de Turin." *RIE, 35* (1898), 334-40, 429-37. Trans. in: U.S., Office of Ed., *Report* (1897-98), *2,* 1452-60.

Hahn, H. "Die neue Bewegung unter den Physiklehrern in den Vereinigten Staaten." *Zeitschrift für den physikalischen und chemischen Unterricht, 20* (1907), 189-95, 259-63.

Hahn, Otto. *A Scientific Autobiography.* New York: Scribner's, 1966.

——. *Vom Radiothor zur Uranspaltung. Eine wissenschaftliche Selbstbiographie.* Braunschweig: Vieweg, 1962.

Hahn-Machenheimer, Hermann. "Die Geryk-Luftpumpe: Patent Fleuss." *Zeitschrift für den physikalischen und chemischen Unterricht, 14* (1901), 285-87.

Haines, George. *Essays on German Influence upon English Education and Science, 1850-1919.* Connecticut College Monograph No. 9. New London: Connecticut College, 1969.

Halle. Universität. *Amtliches Verzeichnis des Personals und der Studierenden* [Semiannual.]

——. ——. *Chronik.* [Annual.]

——. ——. See: Gebhardt, W.; Schrader, W.; *ZBBV, 11.*

Hallock, William. "The Phoenix Physical Laboratories." *Columbia University Quarterly, 5* (1903), 293-96.

Hamburg. *Hamburg in naturwissenschaftlicher und medizinischer Beziehung: den Teilnehmern der 73. Versammlung Deutscher Naturforscher und Ärzte als Festgabe gewidmet.* Hamburg: Voss, 1901.

——. Staatslaboratorium. See: Melle, W. von; Sieveking, H.T.; Voller, A.

Handbuch der Architektur. Teil 4: *Entwerfen, Anlagen und Einrichtung der Gebäude.* Halbband 6: *Gebäude für Erziehung, Wissenschaft und Kunst.* Heft 2, a: *Hochschulen, zugehörige und verwandte wissenschaftliche Institute.* I: *Universitäten und Technische Hochschulen. Physikalische und chemische Institute. Mineralogische und geologische, botanische und zoologische Institute.* 2nd edition. Stuttgart: Kröner, 1905.

Handke, F. "Das Physikalische Institut der Handelshochschule Berlin." *Deutsche Mechaniker-Zeitung* (1907), pp. 57-58.

Hannover. Technische Hochschule. *Festschrift zur 125 Jahrfeier der Technischen Hochschule Hannover, 1831-1956.* Hannover: [Technische Hochschule, 1956].

——. ——. *100 Jahre Technische Hochschule Hannover: Festschrift zur Hundertjahrfeier am 25. Juni 1931.* Hannover: Wilhelm Metzig, 1931.

——. ——. *Programm.* [Annual.]

———. ———. See: Precht, J.; *ZBBV, 15.*

Harker, J.A. "The Physical Laboratories of Manchester University." *Nature, 76* (1907), 640–42.

Harris, Seymour E. *Economics of Harvard.* New York: McGraw-Hill, 1970.

Hartmann, Eugen. "Ständige Ausstellung physikalischer Apparate im Neubau des Physikalischen Vereins zu Frankfurt a.M." *Deutsche Mechaniker-Zeitung* (1907), pp. 146–47.

Hartmann, Hans. *Max Planck als Mensch und Denker.* Thun: Ott, 1953.

Harvard University. *A Guide Book to the Grounds and Buildings.* Cambridge, Mass.: Harvard, 1898.

———. *President's Report.* [Annual.]

———. *Treasurer's Report.* [Annual.]

———. See: Goldberg, S.; Harris, S.; Morison, S.E.

Heidelberg. Universität. *Aus der Geschichte der Universität Heidelberg und ihrer Fakultäten.* Heidelberg: Brausdruck, 1961.

———. ———. See: Baden, Stände-Versammlung; Lenard, P.; *Nature, 65;* Ramsauer, C.; Tompert, H.

Heilbron, John L. *H.G.J. Moseley: The Life and Letters of an English Physicist, 1887–1915.* Berkeley: Univ. of Calif. Press, 1974.

Heinig, Karl. "Das chemische Institut der Berliner Universität unter der Leitung von August Wilhelm von Hofmann und Emil Fischer." Berlin. Universität. *Forschen und Wirken* (1960). Vol. 1, 339–57.

Hellmer, Karl. "Geschichte der K.K. Technischen Hochschule in Brünn." *Festschrift der K.K. Technischen Hochschule in Brünn.* Brünn: Rudolf M. Rohner, 1899. Pp. 1–102.

Helsinki. Universitetet. *Redogörlse.* [Triennial.]

———. ———. See: Tallqvist, H.

Henderson, Yandell, and Maurice R. Davie, eds. *Incomes and Living Costs of a University Faculty.* New Haven: Yale Univ. Press, 1928.

Henriques, Pontus. *Skildringar ur kungl. tekniska högskolans historia: Skrifter utgifna med anledning af inflyttningen i de år 1917 färdiga nybyggnaderna.* 2 vols. Stockholm: Norstadt & Söner, 1917–27.

Herneck, Friedrich. "Wiener Physik vor 100 Jahren." *Physikalische Blätter, 17* (1961), 455–61.

Hertz, Heinrich. *Erinnerungen, Briefe, Tagebücher.* Ed. Johanna Hertz. Leipzig: Akad. Verlag, 1927.

Hesse. Landstände. Zweite Kammer. "Besoldungs-Ordnung für die Grossherzoglichen Staatsbeamten." *Verhandlungen.* 30. Landtag (1897–1900), Beilagenbd. 1, Beilage Nr. 192.

Heussi, Jacob. *Der physikalische Apparat. Anschaffung, Behandlung und Gebrauch desselben.* Leipzig: Frohberg, 1875.

Hewett, W.T. *Cornell University: A History.* 2 vols. New York: Universal Publ. Soc., 1905.

Hight, James, and Alice M.F. Candy. *A Short History of the Canterbury College.* Auckland: Whitcombe & Tombs, 1927.

Hirosige, Tetu, and Sigeko Nisio. "Rise and Fall of Various Fields of Physics at the Turn of the Century." *Japanese Studies in the History of Science,* 7 (1968), 93–113.

Hochschul-Nachrichten. Ed. Paul von Salvisberg. Munich, 1898–[1919].

Hoffmann, Walther Gustav, and J.H. Müller. *Das deutsche Volkseinkommen, 1851–1957.* Tübingen: J.C.B. Mohr, 1959.

Holborn, Hajo. *A History of Modern Germany, 1840–1945.* New York: A.A. Knopf, 1969.

Howard, Henry Fraser. *An Account of the Finances of the College of St. John the Evangelist, 1511–1926.* Cambridge: The Univ. Press, 1935.

Howe, Harley E., and Guy E. Grantham. "Seventy Years of Physics at Cornell." Niels Bohr Library, American Inst. of Phys., New York.

Howorth, Muriel. *Pioneer Research on the Atom . . . . The Life Story of Frederick Soddy.* London: New World, 1958.

Hudson, Claude Silbert, and C.M. Garland. *Tests of a Liquid Air Plant.* University of Illinois. Engineering Experiment Station. Bulletin No. 21. Urbana, Ill.: The University, [1908].

Hufbauer, Karl. *The Formation of the German Chemical Community, 1700–1795.* Diss., Univ. of Calif. at Berkeley, 1970. Ann Arbor: University Microfilms, 1970.

———. "Social Support for Chemistry in Germany during the Eighteenth Century: How and Why Did It Change?" *Historical Studies in the Physical Sciences, 3* (1971), 205–31.

Hull, Callie, et al., comps. *Funds Available in the U.S. for the Support and Encouragement of Research in Science and its Technologies.* Bulletin of the National Research Council, No. 95. Washington, D.C.: NRC, 1921. [1st ed.; 3rd ed., 1934.]

Illinois. University. *Catalogue.* [Annual.]

———. ———. "Note on the History of the Work in Physics at the University of Illinois." *Some Contributions from the Laboratory of Physics of the University of Illinois, Urbana, 1912–1914.* Pp. 5–9.

———. ———. *Trustees' Report.* [Annual.]

———. ———. See: Carman, A.P.; Hudson, C.S.; Nevins, A.

Ingersoll, Leonard R. "The First Hundred Years of the Department of Physics of the University of Wisconsin." Niels Bohr Library, American Institute of Physics, New York.

Innsbruck. Universität. *Uebersicht der akademischen Behörden, Professoren, Privatdozenten, Lehrer, Beamten, etc.* [Annual.]

Iowa. University. See: Stewart, G.W.

Isaachsen, D. "Fysikken og meteorologien." Oslo. Univ. *Det kongelige universitet 1811-1911, 2* (1911), 480-509.

Istoriia estestvoznaniia: literatura opublikovannaia v SSSR, 1917-1956. 3 vols. Moscow: Institut istorii estestvoznaniia i tekhniki, Akademii nauk SSR, 1948-57.

Italy. Istituto centrale di statistica. *Sommario di statistiche storiche dell'Italia, 1861-1965.* Rome: Istituto centrale, 1968.

____. Ministero della pubblica istruzione. *Annuario.*

____. ____. *Annuario generale universitario.* Rome, 1903.

____. ____. *Bolletino ufficiale.* [Annual.]

____. ____. "Legge 19 Luglio 1909, n. 496–Provvedimenti per l'istruzione superiore." Pisa. Università. *Annuario* (1909-10). Pp. 277-336.

____. ____. *Monografie delle università e degli istituti superiori.* 2 vols. Rome: Tipografia operaia cooperativa, 1911-13.

____. Ministero dell'educazione nazionale. Direzione generale delle accademie delle bibliotecha degli affari generali e del personale. *Accademie e istituti di cultura: Fondazioni e premi.* Rome: Palombi, 1940.

Jaeger, Gustav. "Die Lehrkanzeln der Physik und ihre Sammlungen." Vienna. Technische Hochschule. *1815-1915 Gedenkschrift* (1915), pp. 389-99.

Janet, Paul. "La Vie et les oeuvres de E. Mascart." *Revue générale des sciences, 20* (1909), 574-93.

Japan. Imperial Ministry of Education. [Nihon Teikoku Monbusho.] *Annual Report.* [*Nen Pō.*]

Jaubert, Jean. "Enquête sur les industries chimiques françaises." *Revue scientifique, 3* (1905), 97-128.

Jena. Universität. *Amtliches Verzeichnis der Lehrer, Behörden, Beamten, und Studierenden.* [Annual.]

____. ____. *Geschichte der Universität Jena 1548/58-1958.* 2 vols. Jena: Fischer, 1958.

____. ____. See: Schomerus, F.; Wuttig, E.

Jesse, R.H. "Impressions of German Universities." *Educational Review, 32* (1906), 433–44.

Joffe, Abram F. *Begegnungen mit Physikern.* Trans. K. Werner. Leipzig: Teubner, 1967.

———. [Prerevolutionary Physics in Russia and Soviet Physics.] *Uspekhi fizicheskikh nauk, 33* (1947), 453–68.

Johns Hopkins University. *President's Report.* [Annual.]

———. *Register.* [Annual.]

———. See: French, J.C.; Ryan, W.C.

Johnstone, J.H.L. "A Short History of the Physics Department, Dalhousie University, 1838–1956." Halifax, N.S., 1971. Niels Bohr Library, American Institute of Physics, New York.

Joncich, Geraldine. "Scientists and the Schools of the Nineteenth Century: The Case of American Physicists." *American Quarterly, 18* (1966), 667–85.

Jones, Theodore F. *New York University 1832–1932.* New York: NYU Press, 1933.

Junk, Carl. "Physikalische Institute." *Handbuch der Architektur,* Teil 4, Halbband 6, Heft 2, a, I. 2nd ed. (1905), 164–236.

Kalinka, Ernst. "Österreichische Forschungsinstitute: Entwurf einer Rektoratsrede." Innsbruck. Universität. *Bericht über das Studienjahr 1909–10* (Innsbruck, 1911), 38–85.

Kamerlingh Onnes, H. "On the cryogenic laboratory at Leiden and on the production of very low temperatures." Leiden. University. Physical Laboratory. *Communications*, No. 14 (1894).

———. "La liquéfaction de l'hélium, avec des notes sur le laboratoire cryogène de Leyde." Leiden. University. Physical Laboratory. *Communications*, Supplement No. 21a to No. 121–32. (Reprinted from *Rapports et communications du premier congrès international du froid, Paris, octobre, 1908, 2*, 101–26.)

Kangro, Hans. *Vorgeschichte des Planckschen Strahlungsgesetzes. Messungen und Theorien der spektralen Energieverteilung bis zur Begründung der Quantenhypothese.* Boethius. Texte und Abhandlungen zur Geschichte der exakten Wissenschaften, XI. Wiesbaden: Steiner, 1970.

Karlsruhe. Technische Hochschule. *Programm.* [Annual.]

———. ———. *Die Technische Hochschule Fridericiana Karlsruhe: Festschrift zur 125-Jahrfeier.* Karlsruhe: Technische Hochschule, 1950.

———. ———. See: Lehmann, O.; Sieveking, H.

Kastner, Richard H. *Die Geschichte der Technischen Hochschule in Wien.* Wien: Österreicher Bundesverlag für Unterricht, Wissenschaft und Kunst, 1965.

Kausch, Oscar. *Die Herstellung, Verwendung und Aufbewahrung von flüssiger Luft.* 2nd ed. Weimar: Steinert, 1905.

Kay, William Alexander. "Recollections of Rutherford. Being the Personal Reminiscences of Lord Rutherford's Laboratory Assistant . . . ." Ed. Sammuel Devons. *Natural Philosopher.* New York: Blaisdell, 1963. Vol. 1, pp. 127–55.

Kayser, Heinrich. "Erinnerungen aus meinem Leben." Bonn, 1936. Libr. Am. Phil. Soc., Philadelphia.

_____ . "Statistik der Spektroskopie." *Physikalische Zeitschrift, 39* (1938), 466–68.

Kayser, Heinrich, and Paul Eversheim. "Das physikalische Institut der Universität Bonn." *Physikalische Zeitschrift, 14* (1913), 1001–08.

Kelbg, Günter, and W.D. Kraeft. "Die Entwicklung der theoretischen Physik in Rostock." Rostock. Univ. *Wissensch. Zeitschr., Math.-Naturwiss. Reihe, 16* (1967), 839–47.

Kevles, Daniel J. *A Social History of Physics in Modern America.* New York: Knopf, in preparation.

_____ . *The Study of Physics in America 1865–1916.* Diss. Princeton, 1964. Ann Arbor: University Microfilms, 1964.

Kiel. Universität. *Chronik.* [Annual.]

_____ . _____ . *Geschichte der Christian-Albrechts-Universität Kiel, 1665–1965.* 5 vols. Neumünster: Wachholtz, 1965–69.

_____ . _____ . See: Schmidt-Schoenbeck, C.; *ZBBV, 23.*

Kikuchi, Dairoku. *Japanese Education: Lectures delivered in the University of London.* London: John Murray, 1909.

Kindleberger, C.P. *Economic Growth in France and Britain, 1851–1950.* Cambridge: Harvard Univ. Press, 1964.

Kirby, H.W. "The Discovery of Actinium." *Isis, 62* (1971), 290–308.

Kirchner, Joachim, and Hans Martin Kirchner. *Das Deutsche Zeitschriftenwesen: Seine Geschichte und seine Probleme.* Teil II: *Vom Wiener Kongress bis zum Ausgange des 19. Jahrhunderts.* Wiesbaden: Harrassowitz, 1962.

Klein, Felix. "Über die Neueinrichtungen für Elektrotechnik und allgemeine Physik an der Universität Göttingen." *Physikalische Zeitschrift, 1* (1899), 143–45.

Klein, Felix, and E. Riecke. *Über angewandte Mathematik und Physik in ihrer Bedeutung für den Unterricht an den höheren Schulen.* Leipzig: B.G. Teubner, 1900.

Klein, Martin J. *Paul Ehrenfest.* vol. 1: *The Making of a Theoretical Physicist.* Amsterdam: North-Holland Publishing Co.; New York: American Elsevier Publishing Co., 1970.

Kleinwächter, F. "Die Fundirung der Universitäts-Institute in Berlin." *Zentralblatt der Bauverwaltung, 1* (1881), 359–61.

Knott, Cargill G. *Life and Scientific Work of Peter Guthrie Tait.* Cambridge: Cambr. Univ. Press, 1911.

Koch, K. Richard. "Der Neubau des physikalischen Instituts der Technischen Hochschule Stuttgart." *Physikalische Zeitschrift, 12* (1911), 818–31.

Koenigs, Gabriel. "Les nouvelles installations du laboratoire de mécanique de la faculté des sciences de Paris." *La Nature, 42* (1914), 321–25.

Königsberg. Universität. *Chronik.* [Annual.]

——. ——. See: Prutz, H.; *ZBBV,* 7.

Koenigsberger, Leo. *Hermann von Helmholtz.* 1906; New York: Dover, 1965.

Kohl, Max, A.G., Chemnitz. *Price List No. 50.* vols. II and III: *Physical Apparatus.* Chemnitz: Max Kohl A.G. [1911].

Kolde, Theodor. *Die Universität Erlangen unter dem Hause Wittelsbach 1810–1910. Festschrift zur Jahrhundertfeier.* Erlangen & Leipzig: Deichert, 1910.

Konen, Heinrich. "Das physikalische Institut." *Geschichte der Rheinischen Friedrich-Wilhelms-Universität zu Bonn, 1818–1933.* Ed. A. Dyroff. Bonn: Cohen, 1933. Vol. 2, 345–55.

——. *Reisebilder von einer Studienreise durch Sternwarten und Laboratorien der Vereinigten Staaten.* Görres-Gesellschaft. II. Vereinsschrift für 1912. Cologne: Bachem, 1912.

——. "Über die Kruppsche Gitteraufstellung." *Zeitschrift für wissenschaftliche Photographie, 1* (1903), 325–42.

——. "Ueber internationale Organisation naturwissenschaftlicher Forschung." *Sechs Vorträge von der Hildesheimer Generalversammlung.* Görres-Gesellschaft, III. Vereinsschrift für 1911. Cologne: Bachem, 1911. Pp. 35–51.

Kreig, Walter. *Materialien zu einer Entwicklungsgeschichte der Bücher-Preise und des Autoren-Honorars vom 15. bis zum 20. Jahrhundert.* Wien, Zürich: Stubenrauch, 1953.

Krüger, Friedrich. "Die Stellung und das Studium der physikalischmathematischen Wissenschaften an den deutschen Technischen Hochschulen." *Zeitschrift für technische Physik, 2* (1921), 113–21.

Küchler, G.W. "Physical Laboratories in Germany." Office of the Director General of Education in India. *Occasional Reports*, No. 4 (1906), 181–211.

Kuznets, Simon. "National Income." *Encyclopedia of the Social Sciences, 11* (1937), 205–24.

———. *Modern Economic Growth: Rate, Structure and Spread.* New Haven: Yale Univ. Press, 1966.

Kydriavtsev, P.S., O.A. Lezhnev, and A.E. Medunin. ["Physics."] *Istoriia Estestvoznaniia v. Rossii.* vol. 2: *Fizika-Matematicheskie i Khimicheskie Nauk, 1850–1917.* Ed. N.A. Figurovskii. Moscow: Academy of Sciences Press, 1960. Pp. 318–501.

Kyoto. Imperial University. (Kyōto Teikoku Daigaku.) *Calendar.* [Annual.] (In English.)

Lamb, Horace. "Presidential Address (Section A)." British Association for the Advancement of Science. *Report* (1904), 421–31.

Lamprecht, Karl. *Zwei Reden zur Hochschulreform.* Berlin: Weidmannsche Buchhandlung, 1910.

Langdon-Davies, John, ed. *NPL: Jubilee Book of the National Physical Laboratory.* London: H.M. Stationery Office, 1951.

Langevin, André. *Paul Langevin, mon Père.* Paris: Éditeurs Français Réunis, 1971.

Langevin, Paul. "La Physique au Collège de France." Collège de France. *Livre Jubilaire.* Paris: Presses Universitaires, 1932. Pp. 59–80.

———. *Recherches sur les gaz ionisés.* Thèse Paris, 1902. Paris: Gauthier-Villars, 1902.

———. "Pierre Curie." *Revue du mois, 2* (1906), 5–36.

———. "L'Oeuvre de Mascart." *Revue du mois, 8* (1909), 385–406.

Langton, H.H. *Sir John Cunningham McLennan: A Memoir.* Toronto: University of Toronto Press, 1939.

Larmor, Joseph, ed. *Memoir and Scientific Correspondence of the late Sir George Gabriel Stokes.* 2 vols. Cambridge: Cambr. Univ. Press, 1907.

Lausanne. Université. *Catalogue des étudiants.* [Annual.]

———. ———. See: Vaud (canton).

Lauth, Charles. *Rapport général sur l'historique et le fonctionnement de l'École municipale de physique et de chimie industrielles.* Paris: Lahure, 1900.

Lazarev, Petr Petrovich. *Ocherki Istorii Russkoi Nauk*. Moscow-Leningrad: Academy of Sciences Press, 1950.

Lebhardt, Alfred. *Geschichte der Kollegiengebäude der Universität Basel 1460–1936. Festschrift*. Basel: Braus-Riggenbach, 1939.

Lecat, Maurice. *Bibliographie de la relativité*. Brussels: Lamertin, 1924.

Lee, George. *Oliver Heaviside*. London: Longmans, Green, 1947.

Leeds. University. *Calendar*. [Annual.]

——. ——. *Report*. [Annual.]

——. ——. See: Great Britain, Board of Education, *Report; Nature, 78*; Shimmin, A.

Lefavour, Henry. "The Thompson Physical Laboratory at Williams College." *Physical Review, 1* (1894), 451–56.

Lehmann, Otto. *Geschichte des Physikalischen Instituts der Technischen Hochschule Karlsruhe. Festgabe* .... Karlsruhe: G. Braun, 1911.

Leide, Arvid. *Fysiska institutionen vid Lunds universitet*. Lund: Gleerups, 1968. (Published as *Acta universitatis lundensis*. Sectio 1. Theologica, juridica, humaniora. 8.)

——. "Janne Rydberg och hans kamp för professuren." *Kosmos. Fysiska Uppsatser, 32* (1954), 15–32.

Leiden. Universiteit. *Jaarboek*.

——. ——. Natuurkundig Laboratorium. *Het Natuurkundig Laboratorium der Rijks-Universiteit te Leiden in de Jaren 1882–1904. Gedenkboek aangeboden an H. Kamerlingh Onnes, 10 Juli 1904*. Leiden: Ijdo, 1904.

——. ——. ——. *Het Natuurkundig Laboratorium der Rijksuniversiteit te Leiden in de Jaren 1904–1922. Gedenkboek aangeboden aan H. Kamerlingh Onnes*. Leiden: Ijdo, 1922.

——. ——. ——. See: Kamerlingh Onnes, H.; Mathias, E.; *Nature, 54*.

Leipzig. Universität. *Festschrift zur Feier des 500 jährigen Bestehens der Universität Leipzig*. vol. 4: *Die Institute und Seminare der Philosophischen Fakultät*. Part 2. *Die mathematisch-naturwissenschaftliche Sektion*. Leipzig: S. Hirzel, 1909.

——. ——. *Personal-Verzeichnis*. [Semiannual.]

——. ——. See: Des Coudres, Th.; Lilienfeld, J.E.; Wiener, O.

Leiss, C. *Die optischen Instrumente der Firma R. Fuess*. Leipzig: Engelmann, 1899.

Le Maistre, A. *L'Institut de France et nos grands établissements scientifiques*. Paris: Hachette, 1896.

Lenard, Philipp, and Carl Ramsauer. "Tätigkeitsbericht des Radiologischen Instituts an der Universität Heidelberg." *Elektrotechnische Zeitschrift, 31* (1910), 1015–17; *33* (1912), 1103–05; *35* (1914), 1125–27.

Lenz, Friedrich. *Beiträge zur Universitätsstatistik.* Halle (Saale): Waisenhaus, 1912.

Lenz, Max. *Geschichte der Königlichen Friedrich-Wilhelms-Universität zu Berlin.* 4 vols. Halle (Saale): Waisenhaus, 1910–18.

Leobner, H. "Das eidgenössische Polytechnikum in Zürich." *Zeitschrift des Vereins deutscher Ingenieure, 40* (1896), 745–49.

Levy, Maurice. "Allocution . . . séance publique." *Comptes rendus, 131* (1900), 1021–40.

Lexis, Wilhelm. *A General View of the History and Organisation of Public Education in the German Empire.* Trans. G.J. Tamson. Berlin: A. Asher, 1904.

———. "Die neuen französischen Universitäten." *Hochschul-Nachrichten, 11* (1901), 169–76, 193–97, 217–21.

———, ed. *Das Unterrichtswesen im Deutschen Reich. Aus Anlass der Weltausstellung in St. Louis.* vol. 1: *Die Universitäten im Deutschen Reich.* Berlin: Behrend, 1904. vol. 4: *Das Technische Unterrichtswesen.* Part 1: *Die Technischen Hochschulen im Deutschen Reich.* Berlin: A. Asher, 1904.

Liard, Louis. "Les bienfaiteurs de l'Université de Paris." *La revue de Paris, 20,* pt. 2 (1913), 325–46.

———. *L'Enseignement supérieur en France, 1789–1893.* 2 vols. Paris: Armand Colin, 1894.

———. "La Loi sur les universités." *RIE, 32* (1896), 171–74.

———. *L'Université de Paris.* Ed. H. Laurens. Paris: Renouard, 1909.

Liège. Université. *Programme des cours.* [Annual.]

Lilienfeld, J.E. "Das Laboratorium für tiefe Temperaturen (Luft- und Wasserstoffverflüssigung) des Physikalischen Instituts der Universität Leipzig." *Zeitschrift für komprimierte und flüssige Gase, 13* (1911), 165–80, 185–93.

Lille. Université. See: Moissan, H.; Nichols, E.L.

Linders, F.J., and F.E. Lander. *En statistisk undersökning rörande vid Uppsala universitet under åren 1909–1934 avlagda filosofiska ämbetsexamina, särskilt rörande studietidens längd.* Uppsala. Universitet. *Årsskrift* (1936), No. 9. Uppsala: Almqvist & Wiksell, 1936.

Lippmann, Gabriel. "L'Industrie et les universités." Association Française pour l'Avancement des Sciences. *Comptes rendus, 35,* part 1 (1906), 35–42.

Liverpool. University. *Calendar.* [Annual.]

———. ———. See: Great Britain, Board of Education, *Report; Nature, 71.*

Lodge, Oliver. "An Ideal Physical Laboratory for a College." *The Electrician, 26* (1890), 32–33, 66–68.

——. *Past Years: An Autobiography.* London: Hodden & Stoughton, 1931.

Lommel, E. von. "Das neue physikalische Institut der Universität München." *Academische Revue, 1* (1894–95), 261–65.

London. Imperial College of Science and Technology. *Calendar.* [Annual.]

——. Institute of Physics. *The Design of Physics Research Laboratories: A Symposium.* London: Chapman and Hall, 1959.

——. Royal College of Science. *Prospectus.* [Annual.]

——. ——. See: Fowler, A.; Great Britain, Parliament, House of Commons; Tilden, W.A.

——. Royal Institution. *Record of the Royal Institution, 1939.* London: W. Clowes & Son Ltd., 1939.

——. ——. See: Cory, R.; *Nature, 54.*

——. University. See: Bellot, H.H.; Great Britain, Board of Education, *Report*; Smith, T.R.

——. ——. Kings College. *Calendar.* [Annual.]

——. ——. University College. *Calendar.* [Annual.]

Lorey, Wilhelm. "Die Physik an der Universität Giessen im 19. Jahrhundert." *Nachrichten der Giessener Hochschulgesellschaft, 15* (1941), 80–132.

——. *Das Studium der Mathematik an den Deutschen Universitäten seit Anfang des 19. Jahrhunderts.* Abhandlungen über den mathematischen Unterricht in Deutschland, Band 3, Heft 9. Leipzig and Berlin: Teubner, 1916.

Lossen, Wilhelm. *Der Anteil der Katholiken am akademischen Lehramte in Preussen: Nach statistischen Untersuchungen.* Görres-Gesellschaft, 2. Vereinsschrift für 1901. Cologne: P.J. Bachem, 1901.

Lot, Ferdinand. *De la situation faite à l'enseignement supérieure en France.* Cahiers de la quinzaine, série 7, nos. 9, 11. Paris, 1905.

Lot, Fernand. *Jean Perrin et les Atomes.* Savants du monde entier, vol. 16. Paris: Seghers, 1963.

Loudon, James. "The Evolution of the Physical Laboratory." *University of Toronto Monthly, 8* (1907), 42–47.

Louvain. Université Catholique. *Annuaire.*

Lummer, Otto. "Physik." Breslau. Universität. *Festschrift, 2* (1911), 440–48.

Lund. Universitet. *Årsberättelse.* [Annual.] (Bound in Lund. Universitet. *Årsskrift.*)

———. ———. See: Leide, A.; Örtengren, P.; Weibull, J.

Lundbye, Johan Thomas. *Den polytekniske laereanstalt, 1829–1929.* Copenhagen: G.E.C. Gad, 1929.

Lyons. Université. *Année scolaire.*

———. ———. *Université de Lyon, 1900.* Lyons: Université, 1900.

McClenahan, Howard. "The Palmer Physical Laboratory." *Science, 32* (1910), 289–95.

MacKie, John Duncan. *The University of Glasgow, 1451–1951: A Short History.* Glasgow: Jackson, 1954.

MacLeod, Roy M. "The Royal Society and the Government Grant: Notes on the Administration of Scientific Research, 1849–1914." *Historical Journal, 14* (1971), 323–58.

———. "Support of Victorian Science: The Endowment of Research Movement in Great Britain, 1868–1900." *Minerva, 9* (1971), 197–230.

———. "Into the Twentieth Century." *Nature, 224* (1969), 457–61.

———. "Of Medals and Men: A Reward System in Victorian Science 1826–1914." Royal Society of London. *Notes and Records, 26* (1971), 81–105.

———. "Science and the Civil List, 1824–1914." *Technology and Society, 6* (1970), 47–55.

MacLeod, Roy M., and E.K. Andrews. "Selected Science Statistics Relating to Research Endowment and Higher Education, 1850–1914." Science Policy Research Unit, University of Sussex, 1967.

Manchester. University. *Calendar.* [Annual.]

———. ———. *Record of the Jubilee Celebrations at Owens College, Manchester.* Manchester: Sherratt & Hughes, 1902.

———. ———. *The Physical Laboratories of the University of Manchester. A Record of 25 Years' Work.* Manchester: Univ. Press, 1906.

———. ———. See: Great Britain, Board of Education, *Report; Nature, 58, 89*; Steinthal, W.P.

Mandalari, M. "Notizie storiche e descrittive dell'Ateneo e dei suoi istituti." Catania. Università. *Annuario* (1899–1900).

Manegold, Karl-Heinz. *Universität, Technische Hochschule und Industrie: Ein Beitrag zur Emanzipation der Technik im 19. Jahrhundert unter besonderer Berücksichtigung der Bestrebungen Felix Kleins.* Schriften zur Wirtschafts- und Sozialgeschichte, Band 16. Berlin: Duncker and Humblot, 1970.

Mann, C.R. "The Present Condition of Physics Teaching in the United

States." *Broad Lines in Science Teaching.* Ed. Fred Hodson. New York: MacMillan, 1910. Pp. 227-38.

Marburg. Universität. *Catalogus professorum academiae marburgensis: die akademischen Lehrer der Philipps-Universität in Marburg von 1527 bis 1910.* Marburg: Elwert, 1927.

——. ——. *Chronik.* [Annual.]

——. ——. "Der Erweiterungs- und Umbau des mathematisch-physikalischen Instituts der Universität Marburg." *Hessenland. Zeitschrift für Hessische Geschichte und Literatur, 5* (1891), 141-42.

——. ——. *Die Philipps-Universität zu Marburg, 1527-1927.* Marburg: Elwert, 1927.

——. ——. *Verzeichnis des Personals und der Studierenden.* [Semi-annual.]

——. ——. See: Schulze, O.F.A.

Marseille. Université. *Annuaire.*

——. ——. See: du Bourguet, L.

Martin, C. "Quelques chiffres relatifs à la répartition des revenus de la donation Carnegie aux universités écossaises." *RIE, 51* (1906), 18-20.

Martin du Gard, Roger. *Jean Barois.* Paris: Gallimard, 1913.

Massachusetts Institute of Technology. *Catalogue.* [Annual.]

——. *President's Report.* [Annual.]

——. *Treasurer's Report.* [Annual.]

——. See: Goodwin, H.M.

Mathias, E. "Le laboratoire cryogène de Leyde." *Revue générale des sciences,* 7 (1896), 381-90.

Matisse, Georges. *Le Mouvement scientifique contemporain en France.* 4 vol. in 3. Paris: Payot, 1925.

Max-Planck-Gesellschaft. *50 Jahre Kaiser-Wilhelm-Gesellschaft und Max-Planck-Gesellschaft zur Förderung der Wissenschaften, 1911-1961.* Göttingen: Max-Planck-Ges., 1961.

Mayerhöfer, Josef. "Ernst Machs Berufung an die Wiener Universität." *Clio Medica, 2* (1967), 47-55.

Meister, Richard. *Geschichte der Akademie der Wissenschaften in Wien, 1847-1947.* Vienna: A. Holzhausens Nfg., 1947.

Melbourne. University. *Calendar.* [Annual.]

Melle, Werner von. *Dreissig Jahre Hamburger Wissenschaft, 1891-1921.* 2 vols. Hamburg: Broschet-Kommissionsverlag, 1923-24.

Melon, Paul. *L'Enseignement supérieur et l'enseignement technique en France.* 2nd ed. Paris: Armand Colin, 1893.

Mendenhall, Thomas C. "Scientific, Technical and Engineering Education." *Monographs on Education in the United States*. Ed. N.M. Butler. Albany, N.Y.: J.B. Lyon, 1904. Vol. 2, 3-42.

Merz, John T. *A History of European Thought in the Nineteenth Century*. 4 vols. New York: Dover, 1965.

Messina. Università. *Annuario*.

Meyer, A.B. "Studies of the Museums and Kindred Institutions of New York City, Albany, Buffalo and Chicago, with Notes on Some European Institutions." United States National Museum. *Report, 1902-03* (1905), 328-606.

Meyer, Oskar Emil. "Das physikalische Institut der Universität zu Breslau." *Physikalische Zeitschrift, 6* (1905), 194-96.

Meyer, O.E., and K. Mützel. "Über die Störungen physikalischer Beobachtungen durch eine elektrische Strassenbahn." *Elektrotechnische Zeitschrift, 15* (1894), 33-35.

Meyer, Stefan. "Das erste Jahrzehnt des Wiener Instituts für Radiumforschung." *Jahrbuch der Radioaktivität und Elektronik, 17* (1920), 1-29.

———. "Die Vorgeschichte der Gründung und das erste Jahrzehnt des Institutes für Radiumforschung." Vienna. Universität. Institut für Radiumforschung. *Festschrift* (1950), pp. 1-26.

Meyer, Willy. *Die Finanzgeschichte der Universität Zürich von 1833 bis 1933*. Diss. U. Zurich, 1940.

Michigan. University. *Calendar*. [Annual.]

———. ———. *President's Report*. [Annual.]

———. ———. *Treasurer's Report*. [Annual.]

———. ———. See: Dennison, D.M.

Miller, Donald. "Ignored Intellect: Pierre Duhem." *Physics Today, 19* (1966), 47-53.

Miller, Howard S. *Dollars for Research: Science and Its Patrons in Nineteenth-Century America*. Seattle: Univ. of Washington Press, 1970.

———. "Science and Private Agencies." *Science and Society in the United States*. Ed. D.D. van Tassel and M. Hall. Homewood, Ill.: Dorsey, 1966. Pp. 191-221.

Miller, John David. "Rowland and the Nature of Electric Currents." *Isis, 63* (1972), 5-27.

Millikan, Robert A. *The Autobiography of* . . . . London: Macdonald, 1951.

Milne, E.A. *Sir James Jeans: A Biography*. Cambridge: Cambr. Univ. Press, 1952.

Minchin, George M. "Review of Perry, *England's Neglect of Science*." *Nature, 64* (1901), 226–28.

Minerva. Handbuch der gelehrten Welt. Vol. 1: *Die Universitäten und Hochschulen usw., ihre Geschichte und Organisation*. Ed. G. Ludtke and J. Bengel. Strassburg: Karl J. Trübner, 1911.

———. *Jahrbuch der Gelehrten Welt*. Strassburg: Karl J. Trübner, 1891–.

Minnesota. University. *Catalogue*. [Annual.]

———. ———. See: Erikson, H.A.; Gray, J.

Missouri. University. *Catalogue*. [Annual.]

Modena. Università. *Annuario*.

———. ———. See: Mor, C.

Moissan, Henri, and C. Matignon. "Les Nouveaux services et instituts de la faculté des sciences de Lille." *Revue générale des sciences, 6* (1895), 477–93.

Montpellier. Université. *Annuaire de l'université et livret de l'étudiant*.

———. ———. See: Rouzaud, H.

Montreal. McGill University. *Annual Calendar*.

———. ———. *Annual Report of the Governors, Principal and Fellows*.

———. ———. *Formal Opening of the Engineering and Physics Buildings, February 24, 1893*. [Montreal: privately printed, 1893.]

———. ———. See: Eve, A.S.; *Nature, 50*.

Moody, T.W., and J.C. Beckett. *Queen's, Belfast, 1845–1949: The History of a University*. 2 vols. London: Faber and Faber, 1959.

Mor, Carlo Guido. *Storia della Università di Modena*. Modena: Società tip. modenese, 1952.

Morison, Samuel Eliot, ed. *The Development of Harvard University since the Inauguration of President Eliot, 1869–1929*. Cambridge: Harvard Univ. Press, 1930.

Morley, E.W. "Visits to Scientific Institutions of Europe." *American Architect and Building News, 59* (1898), 12–13.

Münster. Universität. *Chronik*. [Annual.]

———. ———. See: *ZBBV, 23*.

Münsterberg, Hugo. "The X-rays." *Science, 3* (1896), 161–63.

Munich. Technische Hochschule. *Die k. k. technische Hochschule zu München: Denkschrift zur Feier ihres 50 jährigen Bestehens*. Munich: Bruckmann, 1917.

———. ———. *Personalstand*. [Semiannual.]

——. ——. *Programm.* [Annual.]
——. Universität. *Amtliches Verzeichnis des Personals, der Lehrer, Beamten, und Studierenden.* [Semiannual.]
——. ——. *Die wissenschaftlichen Anstalten der Ludwig-Maximilians-Universität zu München: Chronik zur Jahrhundertfeier.* Munich: Oldenbourg, 1926.
——. ——. See: Lommel, E. von; Sommerfeld, A.; Wien, W.

Nancy. Université. *Cinquantenaire des facultés des sciences et des lettres, 1854–1904 et séance de rentrée de l'Université de Nancy, 24 novembre 1904.* Nancy: Imprimerie de l'est. 1905.
——. ——. *Livret de l'étudiant.* [Annual.]
——. ——. *Séance de rentrée.* [Annual.]
——. ——. *Université de Nancy, 1572–1934.* Nancy: Pays Lorrain, 1934.
National Research Council. Physics Survey Committee. *Physics in Perspective.* Vol. 1. Washington, D.C.: Nat. Acad. of Sciences, 1972.
[*Nature.*] "Physics and Engineering at the McGill University, Montreal." *Nature, 50* (1894), 558–64.
——. "The London City Companies' Grants to Science and Education." *Nature, 53* (1896), 425–27.
——. "The Davy-Faraday Research Laboratory." *Nature, 54* (1896), 200–01; *55* (1896), 208–09.
——. "The Physical Laboratory at Leiden (Holland)." *Nature, 54* (1896), 345–47.
——. "The New Physical Research Laboratory at the Sorbonne." *Nature, 58* (1898), 12–13.
——. "The New Physical Laboratory of the Owens College, Manchester." *Nature, 58* (1898), 621–22.
——. "Mr. Balfour on Scientific Research." *Nature, 61* (1900), 395–96.
——. "Mr. Balfour on Scientific Progress." *Nature, 62* (1900), 358.
——. "The New Century." *Nature, 63* (1901), 221–24.
——. "The National Physical Laboratory." *Nature, 63* (1901), 300–02.
——. "The Heidelberg Physical Laboratory." *Nature, 65* (1902), 587–90.
——. "The National Physical Laboratory." *Nature, 65* (1902), 466–67, 487–90.
——. "Report on University College." *Nature, 66* (1902), 332–33.
——. "New Buildings of the University of Liverpool: The George Holt Physics Laboratory." *Nature, 71* (1906), 63–65.

——. "The British Science Guild." *Nature, 73* (1905), 10–13.
——. "National Physical Laboratory. Opening of New Buildings for Electrotechnics and Photometry." *Nature, 74* (1906), 205–06.
——. "New Physical and Engineering Departments of the University of Edinburgh." *Nature, 75* (1906), 20–21.
——. "The Needs of the University of Cambridge." *Nature, 75* (1907), 404.
——. "University Needs and the Duty of the State." *Nature, 76* (1907), 35–37.
——. "The Cavendish Laboratory." *Nature, 78* (1908), 152–53.
——. "The New Buildings of the University of Leeds." *Nature, 78* (1908), 257–58.
——. "Scientific Research and the Carnegie Trust." *Nature, 80* (1909), 20–21.
——. "The Extension of the Physical and Electrotechnical Laboratories of the University of Manchester." *Nature, 89* (1912), 46.
——. "The National Physical Laboratory. Opening of New Building." *Nature, 91* (1913), 464–65.
Nebraska. University. *Catalogue.* [Annual.]
——. ——. *Regent's Report.* [Annual.]
Nenot, Henri Paul. *Monographie de la nouvelle Sorbonne.* Paris: Imprimerie nationale, 1903.
——. *La Nouvelle Sorbonne.* Paris: Armand Colin, 1895.
——. "La Nouvelle Sorbonne." *RIE, 29* (1895), 209–44, 326–57, 401–22.
Netherlands. Centraal bureau voor de statistiek. *Statistical Yearbook of the Netherlands, 1972.* Hague: Staatsuitgeverij, 1971.
——. ——. *Zestig jaren statistiek in tijdreeksen, 1899–1959.* Zeist: W. de Haan, 1959.
Neuchâtel (canton). *Décret portant sur le budget de l'état.* [Annual.]
——. *Rapport annuel: enseignement supérieur.*
——. Université. *Autorités, professeurs, étudiants.* [Annual.]
——. ——. *Programme des cours.* [Annual.]
——. ——. *L'Université de Neuchâtel: ses origines, la première et la seconde académie, son organisation actuelle.* Neuchâtel: Université, 1910.
Nevins, Allan. *Illinois.* New York: Oxford Univ. Press, 1917.
Newcastle-upon-Tyne. University of Durham. *Calendar.* [Annual.]
——. ——. See: Fowler, J.T.; Great Britain, Board of Education, *Report.*

Newcomb, Simon. "Conditions Which Discourage Scientific Research in America." *North American Review, 174* (1902), 145–58.

New York. University. Graduate School of Arts and Sciences. *Catalogue.* [Annual.]

New Zealand. University. *Calendar.* [Annual.]

____. ____. See: Hight, J., and Alice M.F. Candy.

Nichols, E.F. "The Wilder Physical Laboratory of Dartmouth College." *Physical Review, 12* (1901), 366–71.

Nichols, E.L. "The New Physics Laboratory at Lille." *Physical Review, 3* (1896), 232–33.

Nitske, Robert W. *The Life of Wilhelm Conrad Röntgen, Discoverer of the X-ray.* Tucson: Univ. of Arizona Press, 1971.

Nitzsche, George E. *University of Pennsylvania. Its History, Traditions, Buildings and Memorials. A Guide for Visitors.* 5th ed. Philadelphia: Winston, 1914.

Northwestern University. *Catalogue.* [Annual.]

____. *Financial Report.* [Annual.]

____. *President's Report.* [Annual.]

Norway. Kirke- og undervisnings-departementet. *Universitets- og skole-annaler.* [Annual.]

____. Statistisk Sentralbyrå. *Historisk Statistikk: 1968: Historical Statistics.* Norges Offisielle Statistikk, ser. xii, Nr. 245. Oslo: Statistisk Sentralbyrå, 1969.

Nottingham. University College. *Calendar.* [Annual.]

____. ____. See: Great Britain, Board of Education, *Report.*

Nye, Mary Jo. *Molecular Reality: A Perspective on the Scientific Work of Jean Perrin.* New York: American Elsevier, 1972.

Örtengren, Per. *Historiska notiser kring Lunds universitets byggnads- och markfrågor.* Lund: Gleerup, 1951. (Lund. Universitet. *Årsskrifter, 46,* No. 5 [1950].)

Ohio State University. *Catalogue.* [Annual.]

____. *President's Report.* [Annual.]

____. *Trustees' Report.* [Annual.]

____. See: Cope, A.

Oldrini, Alexander. "Reform of Superior Education in Italy." U.S. Office of Education. *Report* (1901–02), *1,* 767–87.

Oslo. Universitet. *Det kongelige fredriks universitet 1811–1911: Festskrift.* 2 vols. Kristiania: Forlagt av H. Aschehoug & Co. (W. Nygaard), 1911.

——. ——. "Physics at the University of Oslo, 1900–1910." Personal communication from Egil Spangen and Svend-Erik Svendsen, Matematisk-Naturvitenskapelige Fakultet, Universitetet, Oslo, 21 March 1973.

——. ——. *Universitetet i Oslo, 1911–1961*. 2 vols. Oslo: Universitetsforlaget, 1961.

——. ——. See: Isaachsen, D.; Norway.

Ostwald, Wilhelm. "Naturwissenschaftliche Forschungsanstalten." L. Brauer et al., *Forschungsinstitute* (1930), *1*, 66–92.

Oxford. University. See: Chapman, A.; Ward, W.R.

Padua. Università. *La R. università di Padova e i suoi istituti scientifici. Appendice all'annuario per l'anno accademico 1899–1900 in occasione della esposizione di Parigi.* Padua: Prosperini, 1900.

——. ——. See: Ferraris, C.F.

Palermo. Università. *La R. università di Palermo a S.E. il ministero della istruzione pubblica, all'onorevole giunta parlementare e alle camere dei deputati del senato per i suoi interessi in rapporto all'autonomia universitaria.* Palermo: Virzi, 1899.

Paoli, Louis. "Les Dépenses de l'instruction publique en Italie de 1862 à 1897." *RIE, 43* (1902), 481–502.

Paris. Collège de France. *Annuaire.*

——. ——. *Le Collège de France, 1530–1930.* Paris: Presses Universitaires, 1932.

——. ——. See: Langevin, P.

Paris. Conservatoire national des arts et métiers. Laboratoire d'essais mécaniques, physiques, chimiques, et de machines. *Bulletin, 1* (1903–04).

——. École municipale de physique et de chimie industrielles. See: Lauth, C.

——. École normale supérieure. *Le Centenaire de l'École Normale, 1795–1895.* Paris: Hachette, 1895.

——. ——. See: Peyrefitte, A.; Tannery, J.

——. École polytechnique. *Livre du Centenaire, 1794–1894.* Paris: Gauthier-Villars, 1894–97.

——. ——. See: Callot, J.-P.

——. Exposition universelle, 1900. *Special Catalogue of the Joint Exhibition of German Mechanicians and Opticians.* Berlin: Reichsdruckerei, 1900.

——. Université. Conseil. *Rapport présenté à M. le Ministre de l'Instruction Publique et des Beaux-Arts sur la situation des établissements de l'Université.* [Annual.]

———. ———. Conseil Académique. *Rapports présentés au Conseil Académique sur les travaux et les actes des établissements d'enseignement supérieur.* [Annual.]

———. ———. See: Appell, P.; Berget, A.; Fabry, C.; Koenigs, G.; Liard, L.; Nenot, H.P.

Parma. Università. *Annuario.*

Paschen, Friedrich. "Carl Runge als Spektroskopiker." *Die Naturwissenschaften, 15* (1927), 231-33.

Patterson, Alfred Temple. *The University of Southampton: A Centenary History of the University of Southampton, 1862-1962.* Southampton: Univ. of Southampton, 1962.

Paul, Harry W. "The Issue of Decline in Nineteenth-Century French Science." *French Historical Studies,* 7 (1972), 416-50.

———. "Science and the Catholic Institutes in Nineteenth-Century France." *Societas, 1* (1971), 271-85.

———. *The Sorcerer's Apprentice: The French Scientist's Image of German Science, 1840-1919.* Gainesville: Univ. of Florida Press, 1972.

Paulsen, Friedrich. *An Autobiography.* Trans. and ed. by Theodor Lorenz. New York: Columbia Univ. Press, 1938.

———. *The German Universities and University Study.* Trans. F. Thilly and W.W. Elwang. London: Longmans Green, 1906.

———. *Geschichte des gelehrten Unterrichts auf deutschen Schulen und Universitäten* .... 3rd edition. Edited and continued by R. Lehmann. 2 vols. Berlin and Leipzig: W. de Gruyter, 1921.

Pavia. Università. *Annuario.*

Peixotto, Jessica B. *Getting and Spending at the Professional Standard of Living; a Study of the Costs of Living an Academic Life.* New York: Macmillan, 1927.

Pelletier, Gabriel, and J. Quinet. *Edouard Branly.* Savants du monde entier, vol. 13. Paris: Seghers, 1962.

Pelseneer, Jean. "Historique des Instituts internationaux de Physique et de Chimie Solvay depuis leur fondation jusqu'à la deuxième guerre mondiale." [1962.] Archive for History of Quantum Physics, Berkeley, Copenhagen, and Philadelphia.

Pennsylvania. University. *Catalogue.* [Annual.]

———. ———. *Guide Book 1904-05.* Philadelphia, 1904.

———. ———. *Provost's Report.* [Annual.]

———. ———. See: Nitzsche, G.E.

Perrin, Jean. *Oeuvres scientifiques*. Ed. Francis Perrin. Paris: Centre National de la Recherche Scientifique, 1950.

———. "Faciliter la recherche pour favoriser la découverte" [Speech, 1936]. F. Lot. *Perrin*. 1963. Pp. 162–67.

———. *La Science et l'espérance*. Paris: Presses universitaires de France, 1948.

Perry, John. *England's Neglect of Science*. London: T. Fisher Unwin, 1900.

Perugia. Università. *L'Università di Perugia e i suoi istituti biologici*. (Contains a plan and description of physics institute, pp. 99–101.) Perugia: Tipografia Umbra, 1895.

Peyrefitte, Alain, ed. *Rue d'Ulm: chroniques de la vie normalienne*. Paris: Flammarion, 1963.

Pfetsch, Frank. "Scientific Organization and Science Policy in Imperial Germany, 1871–1914: The Foundations of the Imperial Institute of Physics and Technology." *Minerva, 8* (1970), 557–80.

———. *Zur Entwicklung der Wissenschaftspolitik im Deutschland des 19. und beginnenden 20. Jahrhunderts*. Berlin: Duncker und Humblot, in press.

Picard, Émile. *Discours et mélanges*. Paris: Gauthier-Villars, 1922.

Picavet, François. "L'Enseignement supérieur en Belgique. II: Les Universités de l'État." *RIE, 48* (1904), 206–20.

———. "Dons, donations et legs." *RIE, 49* (1905), 487–513; *50* (1905), 22–48.

Pickering, Edward C. "The Endowment of Research." *Science, 13* (1901), 201–02.

Pisa. Università. *Annuario*.

[Planck, Max.] "Max Planck (1858–1947). Schriftstücke aus seiner amtlichen Tätigkeit (Deutsches Zentralarchiv, Abteilung Merseburg, Rep. 76 Kultusministerium). Zum hundertsten Geburtstage von Max Planck." Berlin: Akademie der Wissenschaften, [1958]. [Documents photostatically reproduced.]

Planck, Max. *Scientific Autobiography and Other Papers*. New York: Greenwood Press, 1968.

Poggendorff, J.C. *Biographisch-literarisches Handwörterbuch zur Geschichte der exacten Wissenschaften*. Vol. 3: *1858–1883;* vol. 4: *1883–1904*. Leipzig: Barth, 1898–1904. Vol. 5: *1904–1922;* vol. 6: *1923–1931*. Berlin: Verlag Chemie, 1924–40.

Pohl, R.W. "Von den Studien- und Assistentenjahren James Francks.

Erinnerungen an das Physikalische Institut der Berliner Universität." *Physikalische Blätter, 28* (1972), 542–44.

Poincaré, Henri. *The Foundations of Science.* Trans. G.B. Halsted. New York: Science Press, 1929.

Poitiers. Université. *Histoire de l'Université de Poitiers passé et présent.* Poitiers: Imprimerie moderne, 1932.

——. ——. *The Institute of Physics at Poitiers University: Its Laboratories and Means of Studies.* Vers l'échange américain. Poitiers: Imprimerie "l'Union," 1919.

Polvani, Giovanni. *Antonio Pacinotti. La vita e l'opera.* 2 vols. Rome: A cura della conf. naz. fascista professionisti ed artisti, anno XII [1934].

Prague. Deutsche Universität. *Die deutsche Karl-Ferdinands-Universität in Prag unter der Regierung seiner Majestät des Kaisers Franz Josef I: Festschrift zur Feier des fünfzigjährigen Regierungsjubiläums.* Prague: Calve, 1899.

——. ——. *Personalstand.* [Annual.]

——. Universitäten. See: Tschermak-Seysenegg, A.

Precht, J. "Die Entwicklung der Physik und die physikalischen Institute." Hannover. Technische Hochschule. *Hundert Jahre* (1931). Pp. 38–42.

Preston, David L. *Science, Society, and the German Jews: 1870–1933.* Diss. Urbana, 1971. Ann Arbor: University Microfilms, 1972.

Previté-Orton, Charles William, comp. *Index to Tripos Lists, 1748–1910, Contained in the Historical Register of Cambridge to the Year 1910.* Cambridge: [Cambridge Univ. Press], 1923.

Priestley, Eliza. "The French and the English Treatment of Research." *Nineteenth Century, 42* (1897), 113–23.

Princeton University. *Bulletin.* [Annual.]

——. *Catalogue.* [Annual.]

——. *Reports of the President and the Treasurer.* [Annual.]

——. See: Cram, R.A.; McClenahan, H.

Pritchett, Henry S. *A Comprehensive Plan of Insurance and Annuities for College Teachers.* Carnegie Foundation for the Advancement of Teaching. *Bulletin* No. 9. New York: CFAT, 1916.

Prokop, August. *Ausbau und Ausgestaltung der k.k. technischen Hochschulen Österreichs: eine Parallele der technischen Hochschulen Österreichs, Deutschlands etc. Aus dem Fest-Vortrage des Rectors der k.k. technischen Hochschule in Wien.* Vienna: Lehmann & Wentzel, 1896.

Prussia. Statistisches Landesamt. *Preussische Statistik.* Vols. 167, 193, 204,

236: *Statistik der preussischen Landesuniversitäten.* Berlin, 1901, 1905, 1908, 1914.

———. ———. *Statistisches Jahrbuch für den preussischen Staat.* Berlin, 1904–.

Prutz, Hans. *Die königliche Albertus-Universität zu Königsberg i. Pr. im neunzehnten Jahrhundert.* Königsberg: Hartung, 1894.

Quantin, A. *L'Exposition du siècle.* Paris: Le Monde moderne, 1900.

[*Radium*]. "Comment on obtient le Radium." *Le Radium, 1* (1904), No. 2, pp. 1, 3–5; No. 3, pp. 5–6; No. 4, pp. 7–8.

Rainoff, T.J. "Wave-like Fluctuations of Creative Productivity in the Development of West European Physics in the Eighteenth and Nineteenth Centuries." *Isis, 12* (1929), 287–319.

Ramsauer, Carl. "Das neue physikalisch-radiologische Institut der Universität Heidelberg." *Elektrotechnische Zeitschrift, 34* (1913), 1364–67.

Ramsay, William. "The Functions of a University." *Nature, 64* (1901), 388–91.

Randall, H.M. "Infrared Spectroscopy at the University of Michigan." *Journal of the Optical Society of America, 44* (1954), 97–103.

Raveau, C. "La vie et l'oeuvre de A. Cornu." *Revue générale des sciences, 14* (1903), 1023–40.

Rayleigh, Robert John Strutt, Fourth Baron. *John William Strutt, Third Baron Rayleigh.* 2nd augmented ed. Madison: Univ. of Wisconsin Press, 1968.

———. *The Life of Sir J.J. Thomson, O.M., Sometime Master of Trinity College, Cambridge.* Cambridge: Cambridge Univ. Press, 1942.

Reading. University. See: Great Britain, Board of Education, *Report.*

Rees, J.K. "Introduction to Catalog, German Association of Mechanicians and Opticians, 1900 Paris Congress." *Science, 12* (1900), 777–85.

Reeves, Floyd W., Frederick J. Kelly, and John D. Russell. *University Plant Facilities.* Chicago: University of Chicago Press, 1933.

Reindl, Maria. *Lehre und Forschung in Mathematik und Naturwissenschaften, insbesondere Astronomie, an der Universität Würzburg von der Gründung bis zum Beginn des 20. Jahrhunderts.* "Quellen und Beiträge zur Geschichte der Univ. Würzburg, Beiheft 1." Neustadt a. d. Aisch: Degener, 1966.

Reingold, Nathan. "American Indifference to Basic Research: A Reappraisal." *Nineteenth-Century American Science: A Reappraisal.* Ed. George Daniels. Evanston: Northwestern Univ. Press, 1972. Pp. 38–62.

———. *Science in Nineteenth-Century America: A Documentary History.* New York: Hill & Wang, 1964.

Rennes. Université. *Annuaire.*

Rezori, W.E. von. "Die neue k.k. Universität in Graz." *Allgemeine Bauzeitung (Wien), 61* (1896), 1-6.

[*RIE*]. *Revue Internationale de l'Enseignement.* [Annual.]

———. "Aperçu sur l'organisation de l'université de Copenhague." *RIE, 26* (1893), 334-41.

———. "Italie: Situation des privatim docentes des universités italiennes." *RIE, 31* (1896), 599-600.

Riecke, Eduard. "Das neue physikalische Institut der Universität Göttingen." *Physikalische Zeitschrift, 6* (1905), 881-92.

Riedler, Alois. *Die Technischen Hochschulen und ihre wissenschaftlichen Bestrebungen.* Berlin: H.S. Hermann, 1899.

Robins, Edward C. *Technical School and College Building, Being a Treatise on . . . Design and Construction.* London: Whittaker, 1887.

Röntgen, Wilhelm Conrad. *Briefe an L. Zehnder.* Ed. L. Zehnder. Zurich: Rascher, 1935.

———. "An die Redaktion." *Physikalische Zeitschrift, 5* (1904), 168.

Rome. Università. *Annuario per l'anno scolastico 1899-1900, con appendice: Compendio di notizie storiche sulla università romana e sugli istituti annessi.* Rome: Fratelli Pallotta, 1900.

Romer, Alfred. *The Discovery of Radioactivity and Transmutation.* New York: Dover, 1964.

Rosa, Edward B. "Plans of the New Buildings for the National Bureau of Standards." *Science, 17* (1903), 129-40.

Rostock. Universität. *Jahresbericht.*

———. ———. *Verzeichnis der Behörden, Lehrer, Beamten, Institute und Studierenden.* [Semiannual.]

———. ———. See: Becherer, G.; Kelbg, G.

Roth, L. "Old Cambridge Days." *American Mathematical Monthly, 78* (1971), 223-36.

Rothblatt, Sheldon. *The Revolution of the Dons: Cambridge and Society in Victorian England.* London: Faber, 1968.

Rouse Ball, W.W. *A History of the Study of Mathematics at Cambridge.* Cambridge: Cambr. Univ. Press, 1889.

Roussy, Baptiste. *Les Progrès de la science et leurs volontaires délaissés: Projet de réorganisation.* Paris: Jules Rousset, 1901.

Rouzaud, Henri. *Les fêtes du VIe centenaire de l'Université de Montpellier.* Montpellier: C. Coulet, 1891.

Royal Society of London. *Catalogue of Scientific Papers, 1800–1900. Subject Index.* Vol. 3: *Physics.* 1 vol. in 2. Cambridge: Cambr. Univ. Press, 1912–14.

——. *Obituaries of Deceased Fellows . . . 1898–1904.* London: Harrison and Sons, 1905. Continued, 1905–1932, in the Society's *Proceedings;* 1932–1954, as *Obituary Notices of Fellows;* and, 1955–, as *Biographical Memoirs of Fellows.*

——. *Year-book.* London: Harrison and Sons, 1896–.

Rubens, Heinrich. "Das physikalische Institut." *Geschichte der königlichen Friedrich-Wilhelms-Universität zu Berlin.* Vol. 3: *Wissenschaftliche Anstalten. Spruchkollegium. Statistik.* Ed. Max Lenz. Halle (Saale): Waisenhaus, 1910. Pp. 278–96.

Rumsey, C. "Mathematics and Science at Cambridge." *Nature, 65* (1902), 510–11.

Runge, Carl, and F. Paschen. "Über die Strahlung des Quecksilbers im magnetischen Felde." Akademie der Wissenschaften zu Berlin. *Abhandlungen* (1902).

Runge, Iris. *Carl Runge und sein wissenschaftliches Werk.* Göttingen. Akademie der Wissenschaften. Math.-phys. Klasse. *Abhandlungen,* 3. Folge, Heft 23. Göttingen: Vandenhoek und Ruprecht, 1949.

Rutherford, Ernest. *Collected Papers of Lord Rutherford of Nelson.* Ed. J. Chadwick. 3 vols. London: Allen and Unwin, 1962–1965.

——. *Radioactive Substances and their Radiations.* Cambridge: Cambr. Univ. Press, 1913.

Ryan, W. Carson. *Studies in Early Graduate Education: The Johns Hopkins University, Clark University, the University of Chicago.* New York: Carnegie Foundation, 1939.

Sachse, Arnold. *Friedrich Althoff und sein Werk.* Berlin: Mittler, 1928.

St. Andrews. University. "Abstract of Accounts" and "Statistical Report." [Annual as follows:] Great Britain. Parliament. House of Commons. *Sessional Papers,* 1902, *81*, 359–400; 1903, *53*, 569–610; 1904, *77*, 189–232; 1905, *61*, 315–58; 1906, *92*, 605–50; 1907, *65*, 669–714; 1908, *86*, 943–88; 1909, *69*, 659–703; 1910, *72*, 781–826; 1911, *60*, 157–202.

——. ——. *Calendar.* [Annual.]

——. ——. See: Cant, R.G.

Sallior, P. "La Réconciliation de la science et de l'industrie." *La Nature, 34* (1906), 374–75.

Sanderson, Michael. *The Universities and British Industry, 1850–1970.* London: Routledge and Kegan Paul, 1972.

Sanford, Fernando. "Department of Physics." Stanford University. *Report of the President* (1906–07), pp. 41–42.

Sarell, Moshe. *Levels and Variation of Productivity in 19th-Century Physical Science: Their Measurement and some of their Social, Economic and Educational Correlates.* Diss. Johns Hopkins, 1970. Ann Arbor: University Microfilms, 1971.

Schallreuter, Walter. "Die Geschichte des physikalischen Instituts der Universität Greifswald." Greifswald. Universität. *Festschrift zur 500-Jahrfeier, 2* (1956), 456–62.

Scheffler, Wilhelm. *Die technischen Hochschulen und Bergakademien mit deutscher Vortragssprache: Organisation und Geschichte . . . nach handschriftlichen und gedruckten amtlichen Quellen.* 6th ed. Leipzig: A. Felix, 1893.

Schlawe, Fritz. *Die Briefsammlungen des 19. Jahrhunderts: Bibliographie der Briefausgaben und Gesamtregister der Briefschreiber und Briefempfänger, 1815–1915.* Repertorien zur Deutschen Literaturgeschichte, Band 4. 1 vol. in 2. Stuttgart: Metzler, 1969.

Schmidt-Schönbeck, Charlotte. *300 Jahre Physik und Astronomie an der Kieler Universität.* Kiel: F. Hirt, 1965.

Schmitt, Eduard. "Chemische Institute." *Handbuch der Architektur,* Teil 4, Halbband 6, Heft 2, a, I. 2nd ed. (1905), pp. 236–382.

_____. *Naturwissenschaftliche Institute der Hochschulen und verwandte Anlagen.* Fortschritte auf dem Gebiete der Architektur, Nr. 7: Ergänzungsheft zu Theil IV, Halbband 6, Heft 2 des 'Handbuchs der Architektur.' Darmstadt: Bergsträsser, 1895.

Schneider, Wolfgang, ed. *Die Technische Hochschule Braunschweig.* Berlin: Länderdienst-Verlag, 1963.

Schomerus, Friedrich. *Werden und Wesen der Carl-Zeiss-Stiftung, an der Hand von Briefen und Dokumenten aus der Gründungszeit, 1886–1896.* Gesammelte Abhandlungen von Ernst Abbe, Band 5. 2nd ed. Stuttgart: Fischer, 1955.

Schrader, Wilhelm. *Geschichte der Friedrichs-Universität zu Halle.* 2 vols. Berlin: Dummler, 1894.

Schreiber, Georg. *Deutsche Wissenschaftspolitik von Bismark bis zum Atomwissenschaftler Otto Hahn.* Arbeitsgemeinschaft für Forschung des Landes Nordrhein-Westfalen, Geisteswissenschaften, Heft 6. Cologne, Opladen: Westdeutscher Verlag, 1954.

Schück, Henrik. *Birka.* Uppsala: Almqvist & Wiksell, 1910. (Uppsala. Universitet. *Årsskrift* [1910], No. 8.)

Schück, Henrik, et al., eds. *Nobel: The Man and his Prizes.* Amsterdam: Elsevier, 1962.

Schuster, Arthur. *Biographical Fragments.* London: MacMillan, 1932.

——. [Presidential Address to Section A.] British Association for the Advancement of Science. *Reports* (1892), pp. 627–35.

——. "International Science." *Nature, 74* (1906), 233–37, 256–59.

——. *The Progress of Physics During 33 Years, 1875–1908.* Cambridge: Cambr. Univ. Press, 1911.

Schulze, O.F.A. "Zur Geschichte des Physikalischen Instituts." Marburg. Universität. *1527–1927,* pp. 756–64.

Schwinge, Erich. *Welt und Werkstatt des Forschers.* Wiesbaden: F. Steiner, 1957.

Seabrook, William. *Doctor Wood: Modern Wizard of the Laboratory.* New York: Harcourt, Brace and Co., 1941.

Sheffield. University. *Calendar.* [Annual.]

——. ——. See: Chapman, A.W.; Great Britain, Board of Education, *Report.*

Shimmin, Arnold N. *The University of Leeds: The First Half-Century.* Cambridge: Cambr. Univ. Press, 1954.

Shryock, Richard. "American Indifference to Basic Science in the Nineteenth Century." *Archives internationales d'histoire des sciences, 2* (1948), 50–65.

Sieveking, Heinrich T. *Die Universität Hamburg.* Düsseldorf: F. Lindner, 1930.

Sieveking, Hermann. *Anleitung zu den Übungen im Physikalischen Institut der Technischen Hochschule zu Karlsruhe.* Karlsruhe: Wilhelm Jahraus, 1903.

Singer, Charles, et al. *History of Technology.* Vol 5: *The Late 19th Century.* Oxford: Clarendon Press, 1958.

Smith, T. Roger, et al. "The New Science Laboratories at University College, London." Royal Institute of British Architects. *Journal, 1* (1893–94), 281–308.

Snyder, Carl. "America's Inferior Position in the Scientific World." *North American Review, 174* (1902), 59–72.

Société d'encouragement pour l'industrie nationale, Paris. *Bulletin.*

Société des Lunetiers. *Instruments pour les sciences physiques, acoustique, électricité, optique, météorologie.* [Paris, 1908].

Société genevoise pour la construction d'instruments de physique. *Prix courant des instruments de physique et d'astronomie* . . . . Geneva: Rey et Malavallon, 1896.

Solberg, Winton U. *The University of Illinois 1867-1894.* Urbana: University of Illinois Press, 1968.

Sommerfeld, Arnold. "Das Institut für theoretische Physik." Munich. Universität. *Wiss. Anstalten* (1926), pp. 290-92.

Southhampton. University. See: Great Britain, Board of Education, *Report;* Patterson, A.T.

Spielmann, Wilhelm. *Handbuch der Anstalten und Einrichtungen zur Pflege der Wissenschaft und Kunst in Berlin.* Berlin: Mayer und Müller, 1897.

Stanford University. *Directory of Officers and Students.* [Annual.]

____. *Information Catalogue.* [Annual.]

____. See: Sanford, F.

Steffens, R.J. "The Problem of Vibrations in Laboratories." Royal Institute of British Architects. *The Design of Physics Buildings.* London: RIBA, 1969. Pp. 25-30.

Steinthal, William Paul. "On the Electrical Equipment of the New Physical Laboratories, Owens College." *Electrical Review* (London), *48* (1901), 615-18.

Stewart, G.W. "The New Physics Building at the University of Iowa." *Contributions from the Physical Laboratory of the State University of Iowa, 1,* No. 5 (1912), supplement, 8-15. Trans. in: *Phys. Zs., 14* (1913), 457-63.

Stötzner, Paul. *Das öffentliche Unterrichtswesen Deutschlands in der Gegenwart.* Leipzig: Göschen, 1901.

Stotesbury, Herbert. "An English University." *Popular Science Monthly, 56* (Nov. 1899), 14-25.

Strassburg. Universität. *Amtliches Verzeichnis des Personals und der Studenten.* [Semiannual.]

____. ____. *Festschrift zur Einweihung der Neubauten der Kaiser-Wilhelms-Universität Strassburg.* Strassburg, 1884.

____. ____. See: Alsace-Lorraine, Landesausschuss.

Studenski, Paul. *The Income of Nations.* Part 1: *History.* New York: N. Y. Univ. Press, 1961.

Stuttgart. Technische Hochschule. *Bericht.* [Annual.]

____. ____. *100 Jahre Technische Hochschule Stuttgart. Zur Jubiläumsfeier, 15.-18. Mai 1929.* Stuttgart: Werbehilfe, 1929.

———. ———. *Programm.* [Annual.]

———. ———. See: Koch, K.R.; Württemberg, Landtag.

Stockholm. Tekniska Högskola. *Skrifter utgivna med anledning av inflyttningen i de år 1917 färdiga nybyggnaderna: Historik och beskrivning rörande nybyggnaderna, jämte avhandlingar och uppsatser av högskolans lärare.* Stockholm: Norstedt, 1918.

———. ———. See: Henriques, P.

Sviedrys, Romualdas. "An Analysis of the Cavendish Laboratory, 1874–1914." Congrès international d'histoire des sciences. XIIe, 1968. *Actes.* Paris: Blanchard, 1971. Vol. 11, 123–27.

———. "The Rise of Physical Science at Victorian Cambridge." *Historical Studies in the Physical Sciences, 2* (1970), 127–51.

Sweden. Ministry of Education. ". . . concerning the state of physics in Swedish universities at the turn of the century. . . ." Personal communication from Jens Cavallin, International Secretariat, Royal Ministry of Education, Stockholm, 27 March 1973.

———. Statistiska Centralbyrån. *Historisk statistik för Sverige. Historical statistics of Sweden. Statistiska Översiktstabeller. Utöver i Del I och II Publicerade.* Stockholm: Statistiska Centralbyrån, 1960.

Switzerland. Department des Innern. *Schweizerische Schulstatistik, 1911–12.* Ed. A. Huber, et al. *Text. 1914: Das schweizerische Schulwesen dargestellt nach den gesetzlichen Grundlagen.* Teil 4: *Die Lehrerschaft aller Stufen.* Berne: Francke, 1915.

Sydney. University. *Calendar.* [Annual.]

———. ———. See: Barff, H.E.

Tallqvist, Hjalmar. *Den nya byggnaden för de fysikaliska inrättningarna vid Kejserliga Alexanders-Universitetet i Finland.* Helsinki: Centraltryckeri, 1911.

Tannery, Jules. "Les Licenses et les agrégations d'ordre scientifique." *RIE, 22* (1891), 473–98.

———. "La Réforme des licences d'ordre scientifique et l'École Normale Supérieure." *RIE, 35* (1898), 44–46.

Teddington, Eng. National Physical Laboratory. *Report.* [Annual.]

Thommen, Rudolf. *Die Universität Basel in den Jahren 1884–1913.* Basel: Reinhardt, 1914.

Thompson, J.S., and H.G. Thompson. *Silvanus Phillips Thompson, D.Sc., LL.D., F.R.S.: His Life and Letters.* New York: Dutton, 1920.

Thompson, Silvanus P. *The Life of William Thomson, Baron Kelvin of Largs.* 2 vols. London: MacMillan, 1910.

——. "Address of the President." Physical Society of London. *Proceedings at the Meetings, 17* (1901), 12–25.

——. "Alfred Marie Cornu, 1841–1902." Royal Society of London. *Proceedings, 75* (1905), 184–88.

Thomson, George P. "Frederick Alexander Lindemann, Viscount Cherwell." Royal Society of London. *Biographical Memoirs of Fellows, 4* (1958), 45–71.

Thomson, J.J. "The Growth in Opportunities for Education and Research in Physics." British Association for the Advancement of Science. *The Advancement of Science: 1931*. London: BAAS, 1931.

——. *Recollections and Reflections*. New York: Macmillan, 1937.

Thwing, C.F. "Pension Fund for College Professors." *North American Review, 181* (1905), 722–30.

Tilden, William A. "The Royal College of Science and the University of London." *Nature, 64* (1901), 583–86.

——. *Sir William Ramsay*. London: Macmillan, 1918.

Tokyo. Imperial University [Tokyo Teikoku Daigaku]. *Calendar*. [Annual.] (In English.)

Tompert, Helene. *Lebensformen und Denkweisen der akademischen Welt Heidelbergs im Wilhelminischen Zeitalter*. Historische Studien. Heft 411. Lübeck and Hamburg: Matthiesen, 1969.

Toronto. University. *Annual Report of the President*.

——. ——. *Auditor's Report to the Board of Trustees on Capital and Income Accounts*. [Annual.]

——. ——. *Calendar*. [Annual.]

——. ——. *The University of Toronto and Its Colleges, 1827–1906*. Toronto: Librarian, Univ. of Toronto, 1906.

——. ——. See: Loudon, J.

Toulouse. Université. *Annuaire*.

——. ——. *Rapport annuel du conseil de l'Université*.

——. ——. *L'Université de Toulouse, son passé–son présent*. Toulouse: Edouard Privat, 1929.

True, Frederick W., ed. *A History of the First Half Century of the National Academy of Sciences, 1863–1913*. Washington: Lord Baltimore, 1913.

Tschermak-Seysenegg, Armin. *Die finanz- und baugeschichtliche Entwickelung der deutschen und der tschechischen Universität in Prag seit der Teilung (1883): Denkschrift*. Brünn: F. Irrgang, [1919].

Tübingen. Universität. "Der Neubau des physikalischen Instituts für die kgl.-württemb. Landes-Universität Tübingen." *Deutsche Bauzeitung, 24* (1890), 213.

——. ——. *Personal-Verzeichnis.* [Semiannual.]

——. ——. *Die Universität Tübingen, ihre Institute und Einrichtungen.* Ed. Th. Knapp and Hans Kohler. Düsseldorf: Lindner, 1928.

——. ——. "Die unter die Regierung Seiner Majestät des Königs Karl an der Universität Tübingen errichteten und erweiterten naturwissenschaftlichen und medizinischen Institute." *Festgabe zum fünfundzwanzigjährigen Regierungs-Jubilaeum Seiner Majestät des Königs Karl von Württemberg.* Tübingen: H. Laupp, 1889. Pp. 1-114.

——. ——. See: Württemberg, Landtag.

Turin. Università. *Annali della R. università di Torino dal 1884 al 1898. Sommario storico-statistico.* Turin: Stamperia reale, 1898.

——. ——. *Annuario.*

——. ——. *La R. università di Torino nel 1900: Anno 496° della Fondazione.* Turin: Stamperia reale, 1900.

——. ——. See: Haguenin, E.

Turner, R. Steven. "The Growth of Professorial Research in Prussia, 1818 to 1848–Causes and Context." *Historical Studies in the Physical Sciences, 3* (1971), 137-82.

Udden, J.A. "Research Funds in the United States." *Science, 55* (1922), 51-52.

United States. Bureau of the Census. *Historical Statistics of the United States.* Washington: G.P.O., 1960.

——. National Bureau of Standards. See: Cochrane, R.; Rosa, E.B.

——. Office of Education. *Report of the Commissioner of Education.* [Annual.]

Uppsala. Universitet. *Handlingar beträffande en nybyggnad för den Fysiska Institutionen vid Uppsala Universitet.* Uppsala: Almqvist & Wiksells, 1903.

——. ——. *Redogörelse.* [Annual.] (Bound in Uppsala. Universitet. *Årsskrift.*)

——. ——. *Uppsala Universitet, 1872-1897: Festskrift.* 3 vols. in 1. Uppsala: Akademiska Boktryckeriet, 1897.

——. ——. See: Ångström, K.; Beckman, A., and P. Ohlin; Linders, F.J., and F.E. Lander; Schück, H.

Vaud (canton). Département de l'instruction publique. *Compte rendu.* [Annual.]

Vavilov, S.I. "Fizicheskii Institut Akademii Nauk za 220 Let." *Uspekhi fizicheskhikh nauk, 28,* No. 1 (1946), 1-50.

Die Vereinigten Fabriken für Laboratoriumsbedarf. *Preis-Verzeichniss über Apparate und Geräthschaften . . . der Allgemeinen Chemie. Liste* No. 60. Berlin: privately printed, [1906].

Vesey, Laurence. *The Emergence of the American University: A Study in the Relations between Ideals and Institutions.* Chicago: Univ. of Chicago Press, 1965.

Vienna. See: Austria, Ministerium für Kultus und Unterricht.

——. Akademie der Wissenschaften. *Almanach.* [Annual.]

——. ——. See: Meister, R.

——. Technische Hochschule. *150 Jahre Technische Hochschule in Wien, 1815–1965.* Ed. H. Sequenz. 2 vols. Vienna and New York: Springer, 1965.

——. ——. *Die K. K. Technische Hochschule in Wien, 1815–1915. Gedenkschrift.* Ed. Joseph Neuwirth. Vienna: T. H., 1915.

——. ——. See: Brandstaetter, F.; Gollob, H.; Jaeger, G.; Kastner, R.H.

——. Universität. *Die feierliche Inauguration des Rectors der Wiener Universität für das Studienjahr . . . .* [Annual.]

——. ——. *Uebersicht der akademischen Behörden, Professoren, Privatdocenten, Lehrer, Beamten, etc.* [Annual.]

——. ——. *Die Universität Wien. Ihre Geschichte, ihre Institute und Einrichtungen.* Düsseldorf: Lindner, 1929.

——. ——. See: Austria, Reichsrat, Haus der Abgeordneten.

——. ——. Akademischer Senat. *Geschichte der Wiener Universität von 1848 bis 1898.* Wien: Holder, 1898.

——. ——. Institut für Radiumforschung. "Festschrift des Institutes für Radiumforschung anlässlich seines 40 jährigen Bestandes (1910–1950)." Akademie der Wissenschaften. Vienna. Mathematisch-naturwissenschaftliche Klasse. *Sitzungsberichte,* Abt. IIa, *159* (1950), 1–57.

——. ——. ——. See: Meyer, S.

Vigneron, H. "L'Exposition annuelle de la société de physique." *La Nature, 42* (1914), 369–73.

Visher, Stephen Sargent. *Scientists Starred, 1903–1943, in "American Men of Science."* Baltimore: Johns Hopkins Press, 1947.

Voigt, Woldemar. "Entgegnung." *Physikalische Zeitschrift, 13* (1912), 1093–95.

——. *Physikalische Forschung und Lehre in Deutschland während der letzten hundert Jahre.* Göttingen: Dieter, 1912.

Voller, A. "Das Physikalische Staats-Laboratorium." Hamburg. *Hamburg in naturwiss. Beziehung* (1901), pp. 205–12.

Wachsmuth, Richard. "Das Physikalische Institut." Frankfurt a.M. Physikalischer Verein. *Neubau* (1908), pp. 88–98.

Waentig, Heinrich. *Zur Reform der deutschen Universitäten.* Berlin: Verlag der Grenzboten, 1911.

Warburg, Emil. "Über die Ziele der PTR: Zur Abwehr." *Physikalische Zeitschrift, 13* (1912), 1091–93.

Ward, William R. *Victorian Oxford.* London: Frank Cass and Co., 1965.

Watanabe, Masao. "The Early Influence of American Science on Japan." Congrès international d'histoire des sciences, X^{e}, 1962. *Actes.* Paris: Hermann, 1964. Vol. 1, 197–208.

Weber, H. Friedrich. "Bericht über die Tätigkeit des physikalischen Instituts, Abtheilung von Prof. Weber, im Jahre 1910." (Copy supplied by K. Lüdi-Knecht, Rektorat, Eidgenössische Technische Hochschule, Zurich.)

Weber, Max. "Science as a Vocation." *Essays in Sociology.* New York: Oxford Univ. Press, 1946.

Webster, Arthur Gordon. "Department of Physics." *Clark University 1889–1899. Decennial Celebration.* Worcester, Mass: Clark, 1899. Pp. 85–98.

Wegener, Else. *Alfred Wegener. Tagebücher, Briefe, Erinnerungen.* Wiesbaden: Brockhaus, 1960.

Weibull, Jörgen. *Lunds universitets historia.* Vol. 4: *1868–1968.* Lund: CWK Gleerup, 1968.

Weinberg, Boris. *L'Enseignement pratique de la physique dans 206 laboratoires de l'Europe, de l'Amérique et de l'Australie.* Odessa: Imprimerie "Economique," 1902.

——. "Essai statistique comparé de l'enseignement pratique de physique." *RGS, 11* (1900), 566.

Weiner, Charles. "Science and Higher Education." *Science and Society in the United States.* Ed. D. Van Tassell and M. Hall. Homewood, Ill.: Dorsey Press, 1966.

Weiss, Pierre. "Electro-aimant de grande puissance." Société française de physique. *Bulletin, 35* (1907), 124–40.

——. "Bericht über die Tätigkeit der Allg. Uebungslaboratorien des physikalischen Instituts während des Jahres 1910." (Copy supplied by K. Lüdi-Knecht, Rektorat, Eidgenössische Technische Hochschule, Zurich.)

——. "Les nouveaux laboratoires techniques de l'école polytechnique de Zurich et ceux de nos facultés des sciences." *Revue générale des sciences, 10* (1899), 55–63.

Welch, W.H. "The Evolution of Modern Scientific Laboratories." *Electrician, 37* (1896), 172–73.

Wertheimer, J. "Higher Technical Education in Great Britain and Germany." *Nature, 68* (1903), 274–76.

Wettstein, Richard. "Die naturwissenschaftlichen Forschungsinstitute in Österreich." *Forschungsinstitute.* Ed. L. Brauer et al. (1930). Vol. 1, 461–70.

Wiedemann, Eilhard. *Das neue physikalische Institut der Universität Erlangen.* Leipzig: Barth, 1896.

Wien, Wilhelm. "Das neue physikalische Institut der Universität Giessen." *Physikalische Zeitschrift, 1* (1899), 155–60.

____. "Das Physikalische Institut und das Physikalische Seminar." Munich. Universität. *Wiss. Anstalten* (1926), pp. 207–11.

Wiener, Otto. "Das neue physikalische Institut der Universität Leipzig und Geschichtliches." *Physikalische Zeitschrift,* 7 (1906), 1–14.

____. "Das physikalische Institut." Leipzig. Univ. *Festschrift, 4,* Pt. 2 (1909), 24–60.

Williams, Ernest Edwin. *Made in Germany.* 5th ed. London: William Heinemann, 1897.

Williams, Howard R. *Edward Williams Morley: His Influence on Science in America.* Easton, Pa.: Chem. Education Publ. Co., 1957.

Willstätter, Richard. *From My Life. The Memoirs of. . . .* Ed. A. Stoll. Trans. L.S. Hornig. New York: Benjamin, 1965.

Wind, C.H. "Overzicht van hetgeen door nederlanders in de jaren 1897 en 1898 op natuurkundig gebied is geschreven." Nederlandsch Natuur en Geneeskundig Congress. *Handelingen,* 7 (1899), 91–139. [Prepared biennially, 1893–1909.]

Winstanley, Denys Arthur. *Later Victorian Cambridge.* Cambridge: Cambr. Univ. Press, 1947.

Wisconsin. University. *Catalogue.* [Annual.]

____. ____. *President's Report.* [Annual.]

____. ____. *Regents' Report.* [Annual.]

____. ____. See: Ingersoll, L.

____. ____. Graduate School. *Catalogue.* [Annual.]

The World Almanac and Encyclopedia. New York: Press Publishing Co., 1901.

Württemberg. Landtag. Kammer der Abgeordneten. "Hauptfinanzetat: Departement des Kirchen- und Schulwesens." *Verhandlungen,* 34. Landtag, Beilagenband 1, Heft 6 (for 1899 and 1900); 36. Landtag,

Beilagenband 2, Heft 6 (for 1904 and 1905); 37. Landtag, Beilagenband 5, Heft 6 (for 1909 and 1910).

——. ——. ——. "Begründung einer Exigenz von 125000 Mk. zur Erweiterung des physikalischen Instituts der Universität Tübingen; Begründung einer Forderung zu einem Neubau für das physikalische Institut der Technischen Hochschule in Stuttgart." *Verhandlungen*, 37. Landtag (1907), Beilagenband 1, Heft 15, 16–18.

Würzburg. Universität. *Personalbestand.* [Semiannual.]

——. ——. See: Reindl, M.

——. ——. Institut für Hochschulkunde. *Bestände der Bibliothek.* Würzburg, c. 1968.

Wuttig, Ernst. "Die Carl Zeiss-Stiftung in Jena und ihre Bedeutung für die Forschung." *Forschungsinstitute.* Ed. L. Brauer et al. (1930). Vol. 1, 441–49.

Wylie, Lawrence. "Social Change at the Grass Roots." *In Search of France.* Cambridge, Mass.: Harvard Univ. Press, 1963. Pp. 159–234.

Yale University. *Catalogue.* [Annual.]

——. *President's Report.* [Annual.]

——. *Treasurer's Report.* [Annual.]

——. Graduate School. *Catalogue.* [Annual.]

——. Sheffield Scientific School. *Catalogue.* [Annual.]

ZBBV = *Zentralblatt der Bauverwaltung.* Citations other than those listed below or by author are of the annual extracts from the Prussian *Staatshaushalt.*

——. "Neubau des physikalischen Instituts in Königsberg i. Pr." *ZBBV*, 7 (1887), 13–14.

——. "Das physikalische Institut in Zürich." *ZBBV, 9* (1889), 135–37.

——. "Neubau des physikalischen Instituts für die Universität Halle." *ZBBV, 11* (1891), 17–18.

——. "Das physikalische Institut in Greifswald." *ZBBV, 11* (1891), 419–20.

——. "Erweiterungsbau der technischen Hochschule in Hannover." *ZBBV, 15* (1895), 465.

——. "Hörsaal des physikalischen Instituts in der Technischen Hochschule in Charlottenburg." *ZBBV, 21* (1901), 230–31.

——. "Die neuen Physikalischen Institute der Universitäten in Münster i.W., Breslau und Kiel." *ZBBV, 23* (1903), 144–46, 157–58.

Zeldin, Theodore. "Higher Education in France, 1848–1940." *Journal of Contemporary History, 2*, No. 3 (July 1967), 53–80.

Zurich (canton). *Staatsrechnung.* [Annual.]

———. Erziehungsrat. *Die Universität Zürich 1833–1933 und ihre Vorläufer: Festschrift zur Jahrhundertfeier.* Zurich: Erziehungsdirektion, 1938.

———. Eidgenössische Technische Hochschule. *Eidgenössische Technische Hochschule, 1855–1955: École Polytechnique Fédérale.* Zurich: Neue Zürcher Zeitung, 1955.

———. ———. *Festschrift zur Feier des 50 jährigen Bestehens des eidgenössischen Polytechnikums.*

1. Teil: Oechsli, Wilhelm. *Geschichte der Gründung des Eidgen. Polytechnikums mit einer Übersicht seiner Entwicklung 1855–1905.* Frauenfeld: Huber, 1905.

2. Teil: *Die bauliche Entwicklung Zürichs in Einzeldarstellungen.* Zurich: Zürcher & Furrer, 1905.

———. ———. Personal communication from K. Lüdi-Knecht, Rektorat, ETH, to A. Miller, Schweizerische Zentralstelle für Hochschulwesen, 20 March 1973.

———. ———. See: Bluntschli, F.; Guye, C.-E.; Leobner, H.; Weber, H.F.; Weiss, P.

———. Universität. "Statistical Data; Physik-Institut Jahresrechnung." Personal communication from E. Bänteli, Abteilung Universität, Erziehungsdirektion des Kantons Zürich, 19 April 1973.

———. ———. *Universität Zürich. Festschrift des Regierungsrates zur Einweihung der Neubauten 18. April 1914.* Zurich: Orell Füssli, 1914.

———. ———. *Verzeichnis der Behörden, Lehrer, Anstalten und Studierenden.* [Annual.]

———. ———. See: Meyer, W.

Zwingli, U., and E. Ducret. "Das schweizerische Sozialprodukt 1910 und in früheren Jahren." *Schweizerische Zeitschrift für Volkswirtschaft und Statistik, 100* (1964), 328–68.

Notes on Contributors

PAUL FORMAN is a curator in the National Museum of History and Technology of the Smithsonian Institution, Washington, D.C., with responsibility for historical collections and exhibits in the field of modern physics. His collaborations with John L. Heilbron began on the project *Sources for History of Quantum Physics,* 1961–1964. Most of the work for the present study was done in 1972–1973 while he was a Visiting Fellow of the Institute of International Studies, University of California, Berkeley.

JOHN L. HEILBRON is Professor of History, and Director, Center for History of Science and Technology, Bancroft Library, University of California, Berkeley.

SPENCER WEART is Director of the Center for the History of Physics, American Institute of Physics, New York. During the years 1971–1974 he was a Postdoctoral Fellow in the Department of History, University of California, Berkeley.